International Conference on Innovation in Education, Science and Culture (ICIESC-2017)

Journal of Physics: Conference Series Volume 970

Medan, Indonesia
8 - 9 November 2017

Editors:

Prof. Moltan
Dr. Martina Restuati
Dr. Juniastel Rajagukguk

ISBN: 978-1-5108-6040-7
ISSN: 1742-6588

Printed from e-media with permission by:

Curran Associates, Inc.
57 Morehouse Lane
Red Hook, NY 12571

Some format issues inherent in the e-media version may also appear in this print version.

Printed by Curran Associates, Inc. (2018)

For permission requests, please contact the Institute of Physics
at the address below.

Institute of Physics
Dirac House, Temple Back
Bristol BS1 6BE UK

Phone: 44 1 17 929 7481
Fax: 44 1 17 920 0979

techtracking@iop.org

Additional copies of this publication are available from:

Curran Associates, Inc.
57 Morehouse Lane
Red Hook, NY 12571 USA
Phone: 845-758-0400
Fax: 845-758-2633
Email: curran@proceedings.com
Web: www.proceedings.com

Table of contents

Volume 970

International Conference on Innovation in Education, Science and Culture (ICIESC-2017)

8–9 November 2017, Medan, Indonesia

Accepted papers received: 15 February 2018
Published online: 19 March 2018

Preface

Papers

IOP Conf. Series: Journal of Physics: Conf. Series **970** (2018) 011001 doi:10.1088/1742-6596/970/1/011001

Preface

2017 International Conference on Innovation in Education, Science and Culture (ICIESC with web link: https://iciesc.unimed.ac.id/) was held on November 8 -9, 2017 in Medan City, Indonesia. This conference was organized by Institut of Research of State University of Medan. Participant with concentration field academicians, researchers and postgraduates student from Indonesia and overseas were attended this conference.

ICIESC2017is the first and will be done annually with main aim is to discuss of recent research special for science, technology, science education and culture. Several topics like: Natural Science, Material Science, Physics Education, Biology Education, Chemistry Education, Vocational, Education, Applied Sciences – Computers, Multimedia Technology, Applied Mathematics, Applied Sciences – Information Technology and Applied Sciences – Engineering have been discussed at the ICIESC 2017. ICIESC 2017 received 150 manuscripts, and 34 manuscripts have been accepted by peer review. There were five keynote speakers who came from USA, Philiphine, Thailand (2 persons) and Indonesia. The ICIESC organizer had been use online submission system to receive all abstract, full paper and also communication with authors. All of information include with comment of reviewer can be cheked real time by author.

The Editors

Prof. Motlan, M.Sc.,P.hD
Dr. Martina Restuati
Dr. Juniastel Rajagukguk

IOP Conf. Series: Journal of Physics: Conf. Series **970** (2018) 011001 doi:10.1088/1742-6596/970/1/011001

Table of Content

ICIESC-15	The Simulation Computer Based Learning (SCBL) for Short Circuit Multi Machine Power System Analysis *Maharani Putri, Rahmaniar*
ICIESC-16	PTR, PCR and Energy Resolution Study of GAGG:Ce Scintillator *Pruittipol Limkitjaroenporn, Wiraporn Hongtong, Hong Joo Kim and Jakrapong Kaewkhao*
ICIESC-17	Physical Characteristics of Laboratory Tested Concrete as a Substituion of Gravel on Normal Concrete *Ronald Butar-butar, Suhairiani, Kinanti Wijaya, Nono Sebayang*
ICIESC-18	Improve Outcomes Study subjects Chemistry Teaching and Learning Strategies through independent study with the help of computer-based media *Gulmah Sugiharti*
ICIESC-19	Producer gas production of Indonesian biomass in fixed-bed downdraft gasifier as an alternative fuels for internal combustion engines *J P Simanjuntak, Lisyanto, E Daryanto and B H Tambunan*
ICIESC-20	The Effect Of Additional Detergent In Crude Palm Oil In The Process Of Separation Stearin *Vina Rezekyah Hasibuan, Nur aini, Febriyanti, Salahudin Al Ayubi Pane*
ICIESC-21	In Vitro Assay of Ethanolic Heat Reflux Extract of *Nicotiana tabacum* L. var Virginia Against Nosocomial Bacteria Pathogen *Andri Pramono, Ahmad Fauzantoro, Irma Rizki Hidayati, Arina Hygea, Oktaviani Sandra Puspita, Hikmah Muktamiroh, Kristina Simanjuntak, Misri Gozan*
ICIESC-22	Review Of Implementation In Bunut Shoes Assistance Program In Order Of Micro, Small And Medium Enterprises Economic In Asahan Regency *M. Saleh Malawat and M. Umar Maya Putra*
ICIESC-23	The Antifungal Inhibitory Concentration Effectiveness Test From Ethanol Seed Arabica Coffee (*Coffea arabica*) Extract Against The Growth Of *Candida albicans Patient Isolate With In Vitro Method, Adam Satria Rakatama, Andri Pramono, Retno Yulianti*
ICIESC-24	A Study of Work Based Learning For Construction Building Workers *Syafiatun Siregar*
ICISC-26	Detection System of Sound Noise Level (SNL) Based on Condenser Microphone Sensor *Juniastel Rajagukguk, Nurdieni Eka Sari*
ICIESC-27	The Implementation of E-learning System at UIN Sumatera Utara in Response to Technology Challenge in Education *Retno Sayekti*
ICIESC-28	Cloud Computing Security Model with Combination of Data Encryption Standard Algorithm (DES) and Least Significant Bit (LSB) *M Basri, H Mawengkang, E M Zamzami*
ICIESC-30	Development of Mathematics Learning Strategy Module, Based on Higher

	Order Thinking Skill (Hots) To Improve Mathematic Communication And Self Efficacy On Students Mathematics Department *Ade Andriani, Izwita Dewi, Budi Halomoan*
ICIESC-31	Biology Factual Knowledge at Eleventh Grade of Senior High School Students in Pacitan based on Favorite Schools *I A Yustiana, Paidi, I S Mercuriani*
ICIESC-32	Implementation of Data Mining to Analyze Drug Cases Using C4.5 Decision Tree *Sri Wahyuni*
ICIESC-33	The development of integrated service centre system for professional teachers empowerment in North Sumatera *S. Gultom, M. M. Simanjorang, Z. Muchtar, A. Mansyur*
ICIESC-34	Biology Procedural Knowledge at Eleventh Grade of Senior High School in West Lampung Based on Curriculum *T M Sari, Paidi, I S Mercuriani*

IOP Conf. Series: Journal of Physics: Conf. Series **970** (2018) 011001 doi:10.1088/1742-6596/970/1/011001

Committee of ICIESC

Advisors

Prof. Dr. Syawal Gultom, M.Pd (Rector of Unimed)
Prof. Drs. Motlan, M.Sc.,P.hD (Chairman of Research Institute of Unimed)
Dr. Martina Restuati, M.Si (Secretary of Research Institute of Unimed)

International Scientific Committee

Prof.Dr.Jakrapong Kaewkhao (NPRU-Thailand)
Prof. Dr. Atty. Cecilio D. Duka, Ed.D. (San Sebastian College of Law, Manila - Philipphine)
Prof. Dr. Ruhizan Mohamad Yassin (UKM Malaysia)
Prof. Dr. Nordin Hussin (UKM-Malaysia)
Assoc. Prof. Dr. Azlizan Mat Enh (UKM-Malaysia)
Dr. Mohd Bin Samsudin (UKM-Malaysia)
Prof. Dr. Chan Kim Ling (Malaysia)
Prof. Alvin P Ang (Philipphine)
Piyawut Srichaikul, Ph.D. (Thailand)
Dr. John Alvarez (Australia)
Dr Michelle Henstridge**(Australia)**
Professor Tim Ackland (Australia)
Prof Loretta Baldassar (Australia)
Prof David Barrie (Australia)
Prof Sandor Bartha (Australia)
Professor Dirk Baur (Australia)
Professor Snezhana Abarzhi (Australia)
Prof Loretta Baldassar (Australia)
Professor Yrjö Engeström (Finland)
Professor Todd L. Hutner (USA)
Dr. Chao Cong (Korea)
Dr. Qian Yang (
Prof. Magdalena Pasarica (USA)
Dr. Ellen K Henriksen (Norway)
Dr. M Rodrigues (Portugal)
Dr. P Simeão Carvalho (Portugal)

Conference Chairperson Dr. Juniastel Rajagukguk, M.Si
Secretary Dr. Zulkifli, M.Sn
Trasurer Dr. Hesti Fibriasari, M.Hum
Secretariat Sadelik S.Pd
Rini Massecurity
Roslaini
Nanda Pratiwi, M.Pd
Rahmad Gultom

Papers and Proceedings Dr. Saronom Silaban, M.Pd
Dra. Ani Sutiani,M.Si
Dr. Muhammad Amin,M.Pd
Dr. Kustoro Budiarta, M.Si
Ricky Syaputra ,S.Pd.,M.Sc

IOP Conf. Series: Journal of Physics: Conf. Series **970** (2018) 011001 doi:10.1088/1742-6596/970/1/011001

Program/Event	Ahmad Shafwan Pulungan, S.Pd.,M.Si Rafael S.Pd.,M.Pd
Public relations	Dr. Ratih Baiduri, M.Si Dewi Andriani, S.Pd.,M.Si
Publication and Documentation	Ater Budiman Sinaga, M.Si Satria Mihardi, S.Pd.,M.Si Pandapotan
Facility	Suyono Agus Ismayadi Cecep
Technology Infomation	Jerry S. Pauned, S.Si. Muhammad Said, S.Pd.,M.Si
Accommodation and Transportation	Henry Situmorang, S.Sos Dr. Makmur Sirait, M.Si Dr. Karya Sinulingga
Consumption	Rosmina Purba Rizki Fadhillah
Receptions	Rahma Diana Asmawati

Peer review statement

All papers published in this volume of *Journal of Physics: Conference Series* have been peer reviewed through processes administered by the proceedings Editors. Reviews were conducted by expert referees to the professional and scientific standards expected of a proceedings journal published by IOP Publishing.

IOP Conf. Series: Journal of Physics: Conf. Series **970** (2018) 012001 doi:10.1088/1742-6596/970/1/012001

Development of k-300 concrete mix for earthquake-resistant Housing infrastructure in indonesia

Fahrizal Zulkarnain[123]
[1]Department of Civil Engineering, Faculty of Engineering,
[2]University of Muhammadiyah Sumatera Utara, Medan, Indonesia

[3]fahrizalzulkarnain@umsu.ac.id

Abstract - In determining the strength of K-300 concrete mix that is suitable for earthquake-resistant housing infrastructure, it is necessary to research the materials to be used for proper quality and quantity so that the mixture can be directly applied to the resident's housing, in the quake zone. In the first stage, the examination/sieve analysis of the fine aggregate or sand, and the sieve analysis of the coarse aggregate or gravel will be carried out on the provided sample weighing approximately 40 kilograms. Furthermore, the specific gravity and absorbance of aggregates, the examination of the sludge content of aggregates passing the sieve no. 200, and finally, examination of the weight of the aggregate content. In the second stage, the planned concrete mix by means of the Mix Design K-300 is suitable for use in Indonesia, with implementation steps: Planning of the cement water factor (CWF), Planning of concrete free water (Liters / m^3), Planning of cement quantity, Planning of minimum cement content, Planning of adjusted cement water factor, Planning of estimated aggregate composition, Planning of estimated weight of concrete content, Calculation of composition of concrete mixture, Calculation of mixed correction for various water content. Implementation of the above tests also estimates the correction of moisture content and the need for materials of mixture in kilograms for the K-300 mixture, so that the slump inspection result will be achieved in planned 8-12 cm. In the final stage, a compressive strength test of the K-300 experimental mixture is carried out, and subsequently the composition of the K-300 concrete mixture suitable for one sack of cement of 50 kg is obtained for the foundation of the proper dwelling. The composition is consists of use of Cement, Sand, Gravel, and Water.

1. Introduction

Some researchers have tested in particular to increase the compressive strength of concrete versus normal concrete, as did Zulkarnain et al. (2016), by addition of paper and also foam to increase the compressive strength of concrete. Concrete can be used in several applications such as pavement, building, foundation, pipeline installation, dam, and other civil building infrastructure (Piyamaikongdech, 2007; Zulkarnain and Ramli, 2011). One alternative to the mixture of concrete components is to use lightweight concrete while maintaining the compressive strength of the existing concrete, so this becomes the solution for the existing concrete mixture. Huang's research (2009) indicates that lightweight concrete can be used as structures in residential buildings by the addition of stiff fibers to concrete mixtures. This is to add strength especially on columns and also beams on building structures to withstand loads.The innovation of concrete technology has also been applied in countries such as Malaysia in 2005 with Smart Tunnel project in Kuala Lumpur, using Lightweight Concrete Method (LCM). With this project the floods that inundate Kuala Lumpur and its surrounding areas can be overcome and channeled into the tunnel to further be directed to the sea (Lee and Hung, 2005). Research conducted by Kuehn (2010) in Canada shows that particles of 10 μm in size can be used for concrete mixtures in building structures. It can be argued that the results of the study are close to the results obtained with ash for concrete mixtures, so the basis of that research is well suited for use in disaster areas. Furthermore, a study by Zulkarnain (2011) suggests that concrete mixtures can be increased in strength by adding silica powder, especially for lightweight concrete that can be used for houses in disaster areas with relatively fast processing and does not require a high cost for each building produced. Previous research by Karolina et al. (2014) concluded that the volcanic ash from Mt. Sinabung could be one of the aggregates in the concrete mix. A study by Zulkarnain (2015) on the strength of concrete with the addition of volcanic ash indicates that this addition can increase the strength of the concrete in such a way that it is possible to reconstruct the earthquake-damaged housing around Mount Sinabung. The innovation of the research in 2016-2018 produced a new K-300 concrete mix for earthquake-resistant housing infrastructure in Indonesia, which can be directly applied to the earthquake area. For the area of North Sumatra, for example, this is appropriate for housing around the safe area of Mount Sinabung, both for short-term and long-term.

2. Literature review

Preliminary study by Zulkarnain et al. (2014) said that a mixture of palm oil can also be used for lightweight concrete mixtures. In this research, the increase of compressive strength value of concrete is obtained which can also be used for housing in disaster area. The results have been published in the Journal of Civil Engineering Research 2014. In Proceedings of the 3rd International Conference on the Geohazard Information Zonation and 5th Seminar & Short Course of HASTAG (GIZ 2014-HASTAG 5), p. 90-98, Karolina et al. (2014) presented the results of research on ash of eruption of Mount Sinabung with the following conclusions: (1) It is found that the water cement factor is so high that it affects the compressive strength of the concrete produced. (2) The visual observation results show that the surface of the brick has the same shape as the particle equation, thus the surface of the brick becomes flat. (3) The use of brick material from the ash of Mount Sinabung resulted in the increase of absorption which in the study by 4.142%. (4) Based on the resulting compressive strength, the addition of 10% ash will result in a compressive strength of 211.01 kg/ cm^2 with treatment and presence of SiO_2 dominance in concrete mixture. (5) From the research, it is found that Mount Sinabung ash can be used in the mix for brick making. The use of ash of eruption of mount Sinabung in concrete mix will increase the compressive strength of concrete at 28 days of testing. The results of the compressive strength test at 28 days were 166.90 kg/cm^2 using no mixture (0%), the compressive strength at 28 days was 173.72 kg/cm^2 by 5% mixture, the compressive strength at 28 days was 207.14 kg/cm^2 by 10 % mixture, and the compressive strength at 28 days was 130.97 kg/cm^2 by 15% mixture. Other chemical compositions are: SiO_2 = 74.3%, AL_2O = 3.3%, CAO = 1.79% (Karolina et al., 2014). Zulkarnain F. (2015), through internal research in Muhammadiyah University of North Sumatra (UMSU), has examined the strength and resilience characteristics of concrete mixtures using silica powder for housing development. Thus, the preliminary study for this research is very supportive and can be the basis for research of K-300 concrete mix for earthquake-resistant housing infrastructure in Indonesia.Inspection of materials becomes the decisive factor in increasing the compressive strength of the tested sample. Selection of materials and techniques or ways of mixing is also a matter of priority before samples are tested. Samples will be tested with cube and cylinder molds for each planned age of up to 28 days. From the test results will be obtained a good value and used as a reference for the mixture of concrete in the disaster area.In the last section, the composition of the concrete mix for every 1 sack of 40 kg cement can be determined based on the volume of the mixture

IOP Conf. Series: Journal of Physics: Conf. Series **970** (2018) 012001 doi:10.1088/1742-6596/970/1/012001

for 1 sack of 40 kg cement. Comparison of volume of mixture for 1 sack cement of 40 kg and final composition of cement: sand: gravel: water can be achieved for earthquake-resistant housing infrastructure in Indonesia.

3. Method of implementation

Primary data is data obtained from the results of research conducted in the laboratory which, among others, are:

a) The materials prepared in the study for mixing of concrete, among others, are: sand, aggregate and Bond Crete supplements. Examination of the materials in the preparation of the study are: Sieve analysis or filtration of coarse aggregate, sieve analysis or filtration of fine aggregate or sand, specific gravity and aggregate absorbance, mud content of aggregate or passing sieve no. 200, and the weight of the aggregate content.

Figure 1. Materials for testing

b) Planning of Concrete Mix design using the *DoE* Method that has been adapted to the conditions of Indonesia.

c) Several types of alternatives are used for the preparation and making of cylindrical test objects which include:
   - Planning used for development or construction of residential houses. The material used is Andalas cement, weighing ± 40 kg per sack.
   - The schedule of activities undertaken in this study include: Characteristic examination, Planning of the composition of the mixture, Preparation of experimental mixture at the initial stage of this study (3 cubes of 15 cm x 15 cm, and 3 cylinders of 15 cm in diameter and height of 15 cm), Quality Inspection of Experimental Concrete Mix with 7 days of age.

In general the resume of design of concrete mix is as follows:

1. Data on Ingredients
   - 40 kg Bosowa Cement.
   - Additive Bond Crete.
   - Water at Public Laboratory of Medan.
   - Aggregates as follows:

Table 1. Data on Ingredients

Aggregate examination	Fine aggregate	Coarse aggregate
- Type	Natural sand	Natural gravel
- Gradation	Zone 2 BS	-
- Maximum diameter	-	40 mm

- Specific gravity of SSD	2.58	2.61
- Absorbance	2.17 %	0.93 %
- Mud content	1.69 %	0.52 %
- Weight of content	1426.86 Kg/m^3	1472.47 Kg/m^3

2. Data for Planning:
   1) Quality of K-300 concrete
   2) Planned slump of 6-9 cm
   3) Planned standard deviation of 50 kg/cm^2
   4) The planning method of *DoE* has been adapted to the conditions of Indonesia.

3. Results of Planning:
   Composition of mixture per 1 m^3 of concrete (Aggregate in state of SSD).
   1) Cement = 448.72 kg
   2) Fine aggregate (sand) = 614.70 kg
   3) Coarse aggregate (gravel) = 1141.58 kg
   4) Water = 175.00 Littre

4. Making of Experimental Mixture:

Table 2. Making of Experimental Mixture

Day/Date	**Monday/August 7, 2017**
Site	Civil Engineering Laboratory of Public Polytechnic of Medan
Type of Mixture	K-300
Object Test	3 cubes of 15 x 15 cm and 3 cylinders of ϕ = 15 cm, height of 15 cm
Additive	Bond Crete
Water content of fine aggregate (sand)	5.82 %
Absorbance of fine aggregate (sand)	2.17 %
Water content of coarse aggregate (gravel)	0.72 %
Absorbance of coarse aggregate (gravel)	0.93 %
Slump obtained	8.17 cm

5. The results of examination of the compressive strength of experimental concrete at age 7 days (estimated 28 days)

Table 3. The results of examination

Parameters		7 days
Average compressive strength	Kg/cm^2	369.48
Standard deviation	Kg/cm^2	5.99
Characteristic compressive strength	Kg/cm^2	459.67

6. Recommendations:

IOP Conf. Series: Journal of Physics: Conf. Series **970** (2018) 012001 doi:10.1088/1742-6596/970/1/012001

1) Based on examination of experimental mixtures, the resulting mixture composition according to the planning can meet the required quality.
2) In the implementation is required to conduct supervision of the quality of the materials of mixture and the quality of concrete by way of sampling according to the provisions and conduct testing of samples in the laboratory.
3) For the purpose of correction of moisture content of the mixture, the examination of the moisture content of the aggregate shall be carried out prior to mixing.
4) The composition of the mixture in units of weight (aggregate in state of SSD) can be seen in Appendix 3.
5) The composition of the mixture in volume unit (aggregate in state of SSD) can be seen in Appendices 5a and 5b.

Figure 2. Mixing of ingredients

Figure 3. Slump test process

The compacting process lasts only about 5 seconds; this process is useful to remove various air and cavities that are around the concrete mixture. All concrete mixtures that have good slump value will be inserted into the mold and vibrated. The vibration is applicable to all cube and cylinder molds in this study.

Figure 4. The compaction process uses a vibrating engine on all samples.

4. Results of testing

Design of K-300 Concrete Mix

The data and materials used during the research process are as follows:

a. Quality of concrete : **K-300**
b. Planned slump : (6 – 9) cm
c. Materials

- Fine aggregate : Type = natural sand
  Specific gravity of SSD = 2.58
  Absorbance = 2.17%
  Gradation = zone 2 BS
- Coarse aggregate : Type = natural gravel

Planning Steps:

1. Planning of Cement Water Factor (CWF)

Estimation of compressive strength of concrete (kg/cm^2) with CWF of 0.5 corresponds to type of cement and aggregate using Table 4.

Table 4. Planning of Cement Water Factor (CWF)

Type of Cement	Type of Aggregate	Compressive Strength (kg/cm^2)			
		3 days	7 days	28 days	91 days
Ordinary Portland Cement (Type I)	Natural (Not destroyed)	200	280	400	460
	Crushed stone	230	320	450	530

Based on the graph of the relationship between Compressive Strength with Cement Water Factor, the CWF value of 0.52 is obtained. Maximum value of CWF for this planning is limited to 0.39 so that the CWF value used is 0.39.

2. Planning of Concrete Free Water (Liter/m^3)

Estimates of the amount of free water (kg/m^3) required for different levels of work can be determined using Table 5

Table 5. Planning of Concrete Free Water (Liter/m^3)

Aggregate		Planned Value of Slump (mm)			
Diameter max. (mm)	Type	0 – 10	10 – 30	30 – 60	60 – 180
20	Natural	135	160	180	190
	Crushed stone	170	190	210	225
40	Natural	115	140	160	175
	Crushed stone	155	175	190	205

The rough aggregate used is a natural type with a maximum diameter of 40 mm and a natural fine aggregate. With planned value of slump of 6-9 cm, it is planned to use free water as much as 175.00 liters/m^3 of concrete.

3. Planning of the Amount of Cement

Based on the smallest CFW value and Free Water then the planned use of cement is as much as:

Total Cement = **Error! Reference source not found.** = **Error! Reference source not found.** = 448.72 kg/m^3

4. Maximum Cement Level

IOP Conf. Series: Journal of Physics: Conf. Series **970** (2018) 012001 doi:10.1088/1742-6596/970/1/012001

The minimum cement quantity for this planning is set at 400 kg/m^3. Thus, the amount of cement used is the calculated amount of cement which is equal to 448.72 kg/m^3.

5. Adjusted Cement Water Factor
Since the amount of cement used does not change, the value of CWF does not need to be adjusted so that the CWF value remains at 0.39.

6. Estimated Aggregate Composition
From the graph of the relationship between the maximum size of coarse aggregate (30 mm), the gradation of fine aggregate (Zone 2 BS), planned value of slump (10 ± 2 cm) and CWF = 0.42 then the composition is obtained:

Fine aggregate : Coarse aggregate = 35% : 65%

7. Estimated Weight of Concrete Content
From the graph of the relation between the relatively specific gravity of the dry aggregate of the surface and the amount of free water, it is estimated that the weight of the wet concrete content is 2380 kg/m^3.

8. Calculation of the Composition of the Concrete Mixture
From the above planning steps for the mixture, the composition per mixed m^3 (aggregate in a state of SSD) is obtained as follows:

- Cement : 448.72 kg
- Fine Aggregate (Sand) : 614.70 kg
- Coarse Aggregate (Split) : 1141. 58 kg
- Water : 175.00 liter

9. Calculation of Correction of the Mixture for Various Water Content
The composition of the mixture per m^3 of concrete in step 8 above is based on the assumption that the aggregate used is in the state of SSD (saturated surface dry). For aggregates not in the state of SSD, the correction of the mixture is carried out as follows:

Cement	: Fixed
Fine Aggregate (Sand)	: Sand of SSD + (Water Content of sand – Degree of Saturation of sand) × Sand of SSD
Coarse Aggregate (Split)	: Gravel of SSD + (Water Content of gravel – Degree of Saturation of gravel) × Gravel of SSD
Water	: Water of SSD – Sand Correction – Gravel Correction

With the description of the research road map above, the results we will achieve in the research will be as expected. This has been done by previous researchers, with the same methods and workings, but the number of molds is increased overall, so a better result is obtained to determine the strength of columns and beams on the main housing/settlement structures. And this can be used in the reconstruction of housing in the vicinity of Mount Sinabung, North Sumatra.

The subsequent process after consideration is to test the compressive strength of each specimen, so that the strength of the sample is known. Testing of compressive strength is also performed on the cylinder by using a cover on the top, so that the flat portion is obtained for more accurate data resulting from the test.

Figure 5. Testing of Compressive Strength

5 CONCLUSIONS

Some results are obtained from testing for both cube and cylinder sample forms according to the age of the plan. Based on examination of experimental mixtures, the composition of the mixture as a result of planning can meet the requested quality.In the implementation is required to perform quality control of materials of mixture and quality of concrete by way of sampling according to the provisions and conduct testing of samples in the laboratory. For the purpose of correction of moisture content of the mixture, the examination of the moisture content of the aggregate shall be carried out prior to mixing. The K-300 mix with additive of 1% Bond Crete can be used for earthquake-resistant housing infrastructure in Indonesia. The use of such mixtures for all infrastructures in buildings or housing is in accordance with established provisions.

ACKNOWLEDGMENTS

The research was conducted with the assistance of the Ministry of Research and Technology, Higher Education of the Year 2016/2017 through the Applied Product Research scheme (PPT).

REFERENCES

[1] Huang, Y.J. (2009). Fiber-Reinforced Syntactic Foam. A Dissertation Presented to the Faculty of the Graduate School, University of Southern California. In Partial Fulfilment of the Requirements for the Degree Doctor of Philosophy (Materials Science).

[2] Karolina R, Syahrizal, Putra M.A, Prasetyo T.A. (2014). The Use of Vulcanic Ash of Mount Sinabung Eruption as the Substitution of Fine Agregate in Making Batako (Mass-Produced Brick). Proceedings of the 3rd International Conference on the Geohazard Information Zonation and 5th Seminar & Short Course of HASTAG (GIZ 2014-HASTAG 5), p. 90-98.

[3] Kuehn, S.C, & Froese, D.G. (2010). Tephra from Ice – A Simple Method to Routinely Mount, Polish, and Quantitatively Analyze Sparse Fine Particles. Microsc. Microanal. 16, p. 218-225.

[4] Lee, Y.L, & Hung, Y.T. (2005). Exploitation of Solid Wastes in Foamed Concrete: Challenges Ahead. Dhir RK, Newlands MD, McCarthy A, Editors. Use of Foamed Concrete in Construction. London, Thomas Telford. p. 15-22.

[5] Piyamaikongdech, A. (2007). Ductile Lightweight Concrete for Lightweight Structural Application. M. S. Thesis, The University of Texas at Arlington.

[6] Zulkarnain, F. (2011). Strength and Durability Properties of Lightweight Foamed Concrete for Housing Construction, PhD Thesis, Universiti Sains Malaysia.

[7] Zulkarnain, F. (2011). Strength and Durability Properties of Lightweight Foamed Concrete for Housing Construction, PhD Thesis, Universiti Sains Malaysia.

[8] Zulkarnain, F, dan Ramli, M. (2011). Performance of Foamed Concrete Mix Design with Silica Fume for General Housing Construction,European Journal of Technology and Advanced Engineering Research, Issue 2.
[9] Zulkarnain, F, Sulieman M.Z, Serri E. (2014). The Effect of Mix Design on Mechanical and Thermal Properties Oil Palm Shell (OPS) Lightweight Concrete, Journal of Civil Engineering Research, Vol.4, No. 3-A, 2014.
[10] Zulkarnain, F. (2015). Karakteristik Kekuatan dan Ketahanan Campuran Beton dengan Penambahan Serbuk Silika untuk Pengembangan Perumahan, APB Universitas Muhammadiyah Sumatera Utara, Mei 2015.
[11] Zulkarnain, F, Sulieman M.Z, Fadila R. (2016). The Potential Usgae Paper Fiber Reinforced Foam Concrete (PFRCFC) Wall Paneling System As An Idea Building Material. International Journal Of Advanced Research (IJAR), Vol. 4, Issue 2, 2016.

Subsurface Structure Determination of Geotermal Area in Siogung-ogung Samosir District by Using Magnetic Method

Togi Tampubolon[1], Juniar Hutahaean[1], Suryani N J. Siregar[1]
Department of Physics, State University of Medan, Indonesia

Corresponding autho: topartam@gmail.com

Abstract. Underwater research often uses geomagnets. It is one of the geophysical methods for measuring magnetic field variations. This research was done to identify how the subsurface rock structure is and determine kinds of rock based on its susceptibility value in Siogung-ogung geothermal area, Pangururan, Samosir District. The tool measurement of total magnetic field called Proton Precission Magnetometer, positioning using Global Position System, and north axis determination using geological compass. Data collection was done randomly with total 51 measuring points obtained. Data analysis started with International geomagnetics Reference Field correction to obtain the total magnetic field anomaly. Then, the data analysis of total magnetic anomaly was done by using surfer program 12. To get a magnetic anomaly cross section used Magdc For Windows program. Magnetic measurement results indicated that the variation of magnetic field strength in each point with the lowest magnetic intensity value of 41785.67 nano tesla. The highest magnetic intensity value is 43140,33. From the results of qualitative interpretation, the magnetic anomaly value is at -200.92 to 1154.45 whereas the quantitative interpretive results of model show the existence of degradation and andesitic rocks, with the value of susceptibility

1. Introduction

The Indonesian archipelago locates near with the boundary of the Eurasian and Indo-Australian tectonic plates. The type of boundary between these two plates is convergent. Indo-Australian Plate is a plate that dips under the Eurasian plate (subduction). In addition, to the east, it met 3 tectonic plates at once, namely the Philippines, Pacific, and Indo-Australian plate [1]. Subduction between two plates causes the formation of volcanoes and oceanic troughs. Similarly, the subduction between the Indo-Australian Plate and the Eurasian Plate causes the formation of volcanoes in rows, that is Bukit Barisan on Sumatra Island and volcanoes along Java island, Bali and Lombok, and the ocean trench, that is Java Trough (Sunda) . Tectonic plates continue to move on each other, their movements causing friction with each other to produce big energy that causes magma rise to the surface and volcanic activity increased or the emergence of geothermal sources on the surface. Therefore, Indonesia is in the fire rings area that have enormous geothermal potential.

Population development in Indonesia is accompanied by the increasing of energy demand. So far, energy supplies still utilize energy sources that dominated by fossils such as petroleum, gas and coal. Dependence on fossil energy causes these energy reserves to be depleted, it is necessary to find alternative renewable energy sources. Renewable alternative energy sources that are very potential and it is time to be developed in Indonesia is geothermal. One of the undeveloped geothermal energy sources is geothermal Siogung-ogung, Pangururan District, Samosir District. So far, Geothermal Siogung-ogung is only used as a hot spring water for tourism object [2].

By looking at the potential of this geothermal, it is necessary to conduct comprehensively research so that geothermal energy resources can be utilized better. In this research, the method for identification of subsurface structure is magnetic method. The application of magnetic methods to geothermal exploration is based on the demagnetization of a material due to temperature. The magnetism of a material will decrease as the temperature rises and even disappears when it reaches the Curie temperature. The observation of the magnetic field is based on measuring the magnetic field strength values. Variations in magnetic field strength values are caused by differences of existing rock susceptibility value at the measurement site as an internal factor and the influence of celestial bodies, especially solar activity as an external factor.

2. Research Method

2.1. Measurement

The position of each measuring point is determined by Garmin Extrex GPS including latitude, longitude, and elevation. The strength measurement of the total magnetic field was carried out using two tools, they are Proton Precision Magnetometer (PPM) G-856 AX model at base point and PPM elsec 770 type at 51 point stations. Measurements at the base point are periodical at certain intervals, while the measurements at each station point are random and repeatable for three times.

2.2. Correction

The magnitude of the measured magnetic field at each station point is essentially the contribution of the earth's main magnetic field, the outer magnetic field, and the anomalous magnetic field. To obtain the anomalous magnetic field strength that produced by the rock beneath of each measurement point surface, it is necessary to make correction to the main magnetic field and external magnetic field which is also measured through PPM. The corrections which are done include:

1. Daily correction (diurnal correction) is a correction to remove the influence of external magnetic field mainly from solar activity. Daily correction is obtained from a variation in magnetic field strength over a specified time interval at the base point corresponding to the measuring time at each station point. If there is a decrease in magnetic field strength, it must be added for its decrease to the measurement result at the station point, and vice versa.
2. Topographic correction is a correction to remove magnetic field magnetic field influence due to topographic differences. There is no general rule in this correction, but if it is deemed as unemagnetized, the correction made only refers to the elevation difference, that is 0.015 nT / m for the equatorial area.
3) IGRF Correction is a correction to eliminate the influence of the main magnetic field or Internatinal Geomagnetic Reference Field (IGRF). The price of IGRF is certain for each point (Latitude, Arc, and elevation) on the earth's surface [3].

The final result of the magnetic method is the strength of the anomalous magnetic field after correction with the equation:

$$\Delta H = H_{OBS} \pm \Delta H_{vh} \pm H_T - H_{IGRF} \quad (1)$$

with:

ΔH = Magnetic Anomaly
H_{OBS} = Measured magnetic field price
ΔH_{vh} = Daily variation price
H_T = Topography correction price
H_{IGRF} = IGRF correction price

IOP Conf. Series: Journal of Physics: Conf. Series **970** (2018) 012002 doi:10.1088/1742-6596/970/1/012002

Suceptibility is the degree of the magnetization of an object due to the influence of the magnetic field. To estimate the magnetic susceptibility value at the measurement point, it can be determined from the equation:

$$K = \frac{\Delta H}{H} \quad (2)$$

with: K = Susceptibility of the magnetic field at the point of measurement
ΔH = Magnetic field Anomaly stregth at measurement point (nT)
H = Average magnetic field stregth in the measurement area (nT)

The value of susceptibility is very important for anomalous objects searching, because of its very characteristic for each type of mineral or metal mineral.

2.3. Interpretation

Interpretation of geomagnetic data is done both qualitatively and quantitatively. Qualitative interpretation is done by presenting the magnetic field strength of anomaly in two-dimensional map using Surfer 12. The contour pattern of anomalous magnetic field shows the distribution of geomagnetic objects or geological structures at Earth subsurface. Furthermore, the magnetic field anomaly pattern is interpreted based on local geological information in the form of the distribution of magnetic objects or geological structures, as the basic for estimating the actual geological condition. Quantitative interpretation aims to describe the subsurface structure to determine the lithology of the research area [5]. Interpretation is done to create a geomagnetic sectional model using Mag2dc software. The geomagnetic sectional model is linked to a table of susceptibility values in rock type and geological data of the research area to determine the subsurface rock structure.

3. Results

3.1. Distribution Patterns of Magnetic Anomalies

The distribution pattern of anomalous magnetic field strength in the measurement area is shown in Figure 1 below:

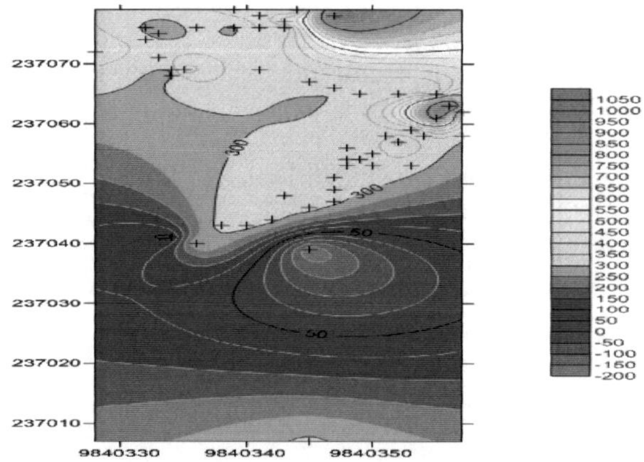

Figure 1. Distribution Patterns of Siogung-ogung Magnetic Anomalies

Based on the picture above, it is found that the highest magnetic field strength is 1154.45 (nT) at A27 point and the lowest anomaly is -200,92 (nT) at A1 point. It is accordance with the survey area where the area is a magnetic measurement point that close to the emergence of hot water. The price of low and medium magnetic anomalies is spread over the west which is closely related to the formation of geothermal manifestations in this area. This strong magnetic field states the effect of rocks demineralization due to high temperatures. This area is interpreted as a geothermal prospect area that needs to be examined more thoroughly to ensure its energy potential [6].

3.2. Susceptibility of Siogung-ogung Geothermal Rock

Based on the magnetic field strength of anomaly, then the susceptibility calculation is using equation 2). The pattern of anomalous magnetic field strength distribution in the measurement area is shown in Figure 2 below:

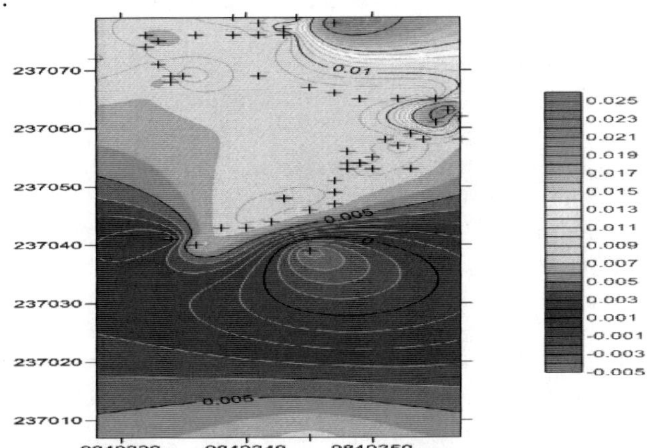

Figure 2. Pattern of Susceptibility Distribution at Siogung-ogung

From the calculation of susceptibility value, it is obtained that Siogung-ogung geothermal region has suscetibility value from the lowest to highest that is: - 0,00479 x 103 until 0,0275 x 103. The susceptibility value obtained will be used to determine the type of rock at subsurface of Siogung-ogung geothermal area through quantitative analysis with modeling using Mag2dc software.

3.3. Magnetic Anomaly Modeling

A quantitative interpretation is made to create a geomagnetic sectional model using Mag2dc software for susceptibility data. Modeling is done by making the line section (AA' path) from the lowest anomaly to the highest that suspected as a source of magnetic anomalies. In numerical modeling, some geometric magnetic field parameters are needed in the research area including IGRF value (41987.1 nT), declination angle (-0.1210 °), inclination angle (-0.0164 °) and other parameters. The modeling results are shown in Figure 3 below:

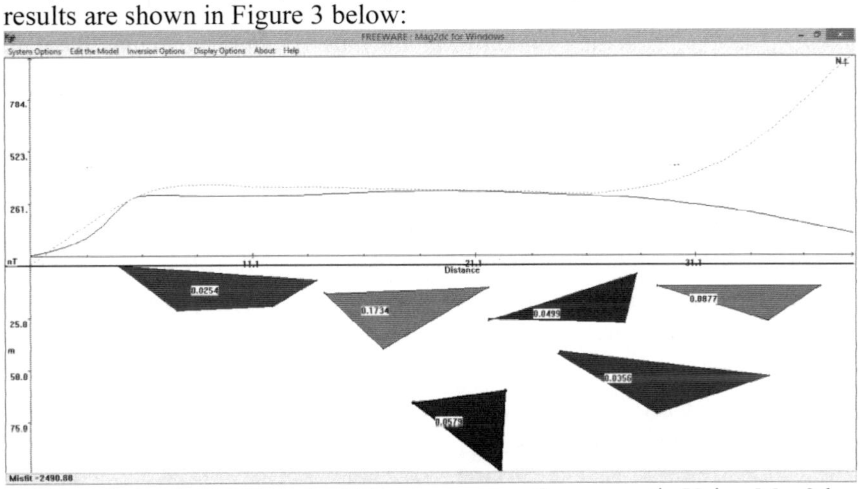

Figure 3. Sectional Model of Siogung-ogung Geomagnetic Using Mag2dc

Figure 3 shows the AA' cross-sectional model, where the x-axis denotes the length of the latitude, the positive y-axis represents the strength value of magnetic anomaly, field and the negative y-axis represents depth. From the modeling results, it can be determined rocks lithology based on the susceptibility value. The top layer with a value of k = -0.0254 is interpreted as pyroclastic rock at a depth of 0 to 20 m which serves as cup rock on a geothermal system. Layers with values of k = 0.0499 are interpreted as andesite rocks at depths of 4 to 25 m which serve as a barrier to the loss of geothermal steam. The layer with the value of k = 0.0877 at a depth of 10 to 25 m, k = 0.1734 at a depth of 10 to 35 m, k = 0.0356 at a depth of 41 to 60 m, and k = 0.0579 at a depth of 60 to 100 m, interpreted as a strong alternated rock or have undergone many changes and decreased magnetism properties due to heat. The base of this layer is a reservoir zone and beneath this layer is the geothermal host rock and geothermal vapor.

The contrast of negative and positive susceptibility values to area A is interpreted as a Horst Graben structure, a fracture used as a geothermal vapor exit and in that area there is a fumarole crater which is a geothermal manifestation. If there it is related with the value of the susceptibility table by Telford [8], the rock is andesite that have a susceptibility of 0.0100 to 0.0500 in line with the geological data of the research area. This result is in line with the previously obtained that andesite rocks have a susceptibility between 0.0200 to 0.0400 [9].

4. Conclusion

From the research results, it can be concluded, such as (1) The magnetic anomaly value in the survey area ranges from - 200.92 nT at coordinates 237039 N 9840345 E to 1154.45 nT at coordinates 237063 N 9840356 E and (2) Based on the value of susceptibility obtained, the types of rocks found in this research area are desiccation and andesitic rocks with the value of susceptibility (0.0254, 0.0499, 0.0877, 0.1734, 0.0356, 0.00579)

References

[1] Badan Geologi, (2010), Hidup di atas tiga lempeng : Energy, Mineral dan Geodiversity, Badan Geologi, Kementerian Energy dan Sumber Daya Mineral, Bandung.

[2] Badan Pusat Statistik, (2015), Statistika Daerah Kabupaten Simalungun 2015, Badan Pusat Statistika Kabupaten Samosir, Medan : Samosirkab.bps.go.id (diakses pada 09 Oktober 2016).

[3] Broto, S., T. T, Putranto, (2011), Aplikasi Metode Geomagnet Dalam Eksplorasi Panas Bumi, UNDIP: Semarang.

[4] http://swdcwww.kugi.kyoto-u.ac.jp/igrf/point/index.html, model field at a point by IGRF accessed, August, 23, 2016.

[5] Winarsih, F. P., (2014), Identifikasi Litologi daerah Manifestasi Panas Bumi Parangwedang Kabupaten Bantul DIY dengan Metode Magnetik, Skripsi Fakultas Sains dan Teknologi, UIN Sunan Kalijaga, Yogyakarta.

[6] Singarimbun, A., Cyrke,A.D.N.B, and Riva,C.F., (2013). Penentuan Struktur Bawah Permukaan Area Panas Bumi Patuha dengan Menggunakan Metode Magnetik, Jurnal Matematika dan Sains, Agustus 2013, Vol. 18 No.2. Sulawesi Utara.

[7] Setiawan, Sigit, (2012). Energi Panas Bumi Dalam Kerangka MP3EI : Analisis Terhadap Prospek, Kendala, danDukunganKebijakan, JurnalEkonomidan Pembangunan, Vol. XX (1).

[8] Telford. W. M., L. P. Geldart dan R. E. Sheriff., (1991), Applied Geophysics, Second Edition. Cambridge University Press: USA.

[9] Lita, F., (2012), Identifikasi Anomali Magnetik di Daerah Prospek Panasbumi Arjuna- Welirang, Skripsi, FMIPA, Universitas Indonesia (UI), Depok.

IOP Conf. Series: Journal of Physics: Conf. Series **970** (2018) 012003 doi:10.1088/1742-6596/970/1/012003

The role of vocational education in science and technology developement in the era of globalisation

Sudirman[12]
[1]Faculty of Economy, Universitas Negeri Gorontalo, Indonesia
[2]sudirman@ung.ac.id

Abstract. Globalization refers to a condition by which a world becomes competitive. In such a world, people are urged to do a job properly since they will be the asset to advance their nation further; this is also to strengthen the nation's capital markets. Both the formal and informal vocational education is proposed to produce well-capable and highly-competitive graduates. This is considered important as one effort to deal with some challenges as well as a rapid advance in science and technology in the globalized world. This present study aims at exploring the contribution of vocational education in response to the development of science and technology in the globalized world. Furthermore, this study is a development of a notion that the vocational education system is proposed as a solution to the highly-competitive situation and challenges due to the globalization. To put it simply, vocational education significantly contributes towards the development of science and technology in a globalized world and how to deal with the situation. Such an education system equips the student with skills needed to become a productive and professional individual. This is also to produce better and competitive human resources that are able to compete with other nations.

1. Introduction

Education is vital in producing high quality, competitive, and professional human resources as the future generation of a nation. In addition, education aims to support the nation's development and maintain the sustainability of the progress in such a competitive and a more complex advancement of science and technology. These are the rationales of the importance of education in the current globalized world. Planning is required in addressing the development of science and technology as well as in producing high quality of human resource. This target can be achieved through education since it is the most efficient and applicable approach for people. It is also possible to produce human resources that are not only knowledgeable but also skillful. Having adequate understanding regarding science and technology as well as mastering a particular skill turn to be the rationale of the educational system evaluation all over the world including Indonesia. Indonesian government continuously work at their best to prepare high-quality resources as the nation's next generation. This is because of the contribution of those people to developing their society, promoting prosperity, improving knowledge, skill, and capacity of people, boosting economic growth and alleviating poverty (Pasban, et al, 2016).Competent human resources serve as the asset of Indonesia in creating a sustainable development and competitive environment throughout the nation. These can only be achieved through a better education system. The atmosphere of competition in all sectors of the globalized world urges Indonesia to be able to get involved with such a situation and reach the top through a productive society. Therefore, the society, especially the further generation, is required to become knowledgeable and tech-savvy as this has been constituted in the 1945 Constitution. The proposed education system should refer to these purposes to enable the society dealing with the 21st century's constant development of science and technology. It is also necessary to design an education system that can improve learners' competence to survive in the globalized world. In other words, the good education system is able to create high-quality human resources in a competitive atmosphere of the globalization. The government is responsible for producing qualified human resources in developing the nation

IOP Conf. Series: Journal of Physics: Conf. Series **970** (2018) 012003 doi:10.1088/1742-6596/970/1/012003

through designing an education system. This system, which refers to the vocational education system, aims to equip the individuals with skills and competitive spirit in business and industry sector.

2. Vocational Education

Vocational education is that part of education which integrates both the theory and practice or job training in related fields. Students are equipped with the knowledge from the regular class and the experience from the hands-on training in the sectors that the students enroll. The act number 20 of 2013 considering the national education system points out that vocational high school (*Sekolah Menengah Kejuruan* or *SMK*) is a secondary education program for preparing learners for a specific job. Although vocational education is considered as an equal program with the senior high school (*Sekolah Menengah Atas* or *SMA*), it has also a number of differences, such as (1) theory-oriented and hands-on-training-oriented, (2) decisions for existence and legitimacy, (3) focus on the content of the curriculum, (4) criteria of a successful learning, (5) aware of the development of a society, and (6) correlation with society in terms of the cooperation (Finch & Crankilton, 1982, p.12-13). The government keeps encouraging the secondary school graduates to continue their study to vocational school to equip the students with competence and skill in a particular field. This is based on the concept by Joyonegoro (1998) which explains that (1) Competent individuals are required for the development of a nation, (2) The globalized world is now becoming more competitive in every sector, (3) Technology development plays a significant role in the advancement of a nation, (4) Skilful individuals are likely to get a job and secure a better future. Vocational education is designed to prepare the youth to promote prosperity, to possess a particular skill, and to gain self-confidence as the citizen of Indonesia. The aim of such an education system is also to prepare the youth as a professional in his or her sector of business or industry. The influence of foreign investment which results in the establishment of new companies or jobs urged a program well-known as Small and Micro Enterprise (*Usaha Kecil Menengah* or *UKM*). This program contributes to the employment growth especially the non-wage earners (Nen, 2015). According to Tatiana et al. (2015), the graduates of the vocational school who do not go to college increase the number of the issue of unemployment around youth. Furthermore, Tatiana et al. (2015) proves that (1) there is a correlation between the number of unemployment issue and socio-economical aspect with the demographics of an area, (2) differentiating the regional-based approach regarding the employment policy which aims at alleviating the unemployment issue, and (3) vocational education plays a major role to solve the jobless people around youth.

Productive and skilled employers, once they get involved with the competitive work environment, are expected to produce brilliant ideas and high quality and competitive goods. Possessing such traits is possible to solve the problems of being jobless since all the companies seek only those who are competent, eager, and knowledgeable regarding the work environment. Individuals with the characteristics mentioned are also required to take the risk to run their private business, provide jobs for other people, and promote a mindset of being an independent and productive person. Professionals are obliged to accomplish their work efficiently. This is based on their disipline, the acknowledgment from the society, and the codes of conduct that act as the standard operating procedure, or SOP in short, of their job. Professional employees possess the following characteristics of the 21st century human resources, namely (a) Critical-Thinking and Problem-Solving Skills, (b) Communication and Collaboration Skills, (c) Critical-Thinking and Problem-Solving Skills, (d).Communication and Collaboration Skills, (e) Creativity and Innovation Skills, (f) Information and Communications Technology Literacy, (g) Contextual Learning Skills, (h) Information and Media Literacy Skills (Rahdiyanta, 2014). Rahdiyanta further classifies the characteristics that represent the 21st century individual, e.g., (a) Leadership, (b) Personal Responsibility, (c) Ethics, (d) People Skills, (e) Adaptability, (f) Self-Direction, (g) Accountability, (h) Social Responsibility, (i) Personal Productivity.

3. The Development of Science and Technology

The constant, uncontrollable development of science and technology has been shifted towards a negative situation in which the advancement becomes threatening, such as the misuse of the recent science and technology for the purpose of war. By that, implementing the four tenets of learning is effective to tackle the issue. These involve (1) learn to know, (2) learn to work, (3) learn to live and to develop, and (4) learn to be individuals in their truest sense. These paradigms of education are expected to be able to cope with the issues regarding the rapid advancement of science and technology which is being more complex currently. On the other hand, the development of science and technology brings good to people in every sector including education. Some examples of the contribution of science and technology to the field of education are (1) The emergence of electronic media can help people distributing information and knowledge. In other words, teachers are not the only one who can disseminate information to others. (2) The emergence of new learning methods and model which ease teachers and students during the class. Through the help of technology, the teachers can design new learning strategies that can help them transfer the knowledge to their students. As a result, the students will have no problems to process the information, (3) Technology provides a new learning method, namely e-learning, in addition to the current mode of delivery.

There are four major components of cloud-technology-based and social-media-based learning which also turn to be the challenge in implementing these models. Those are (1) The tenets of learning model, challenge-based learning through cloud technology and social media, (2) The aims of the instructional model; to develop managerial information skill. (3) Challenge based learning by implementing the internet technology and social media consists of the preparatory stage prior to designing the learning components and instructional, (4) The assessment involves data examination regarding the information management skill from three components, e.g., information search, library skills, and resource evaluation (Yoosomboon, 2014). In response to the development of science and technology, promoting a better education system to prepare the students for success in their work and therefore become a competitive advantage for the organization.

4. Globalization

According to Kopweh (as cited in Saeed, 2015), globalization is the integration of culture, economy, and politics among the nations. Globalization is a condition in which a global market competition between every nation urges competitive and high-quality goods; this requires the support from human resources. Globalization brings challenges to the education sector which Kurniawan (1999) further classifies these into four, such as (1) improving the productivity of the nation for a sustainable development, (2) conducting an extensive and in-depth research on the transformation of the social structures within a society, the shift from the traditional to modern society, and how its impact on the development and quality improvement of human resources, (3)competition that becomes stiffer; this refers to efforts in promoting the competitive environment and producing creative products as the results of the findings, concepts, and skills in science and technology, (4) the emergence of new movements regarding science and technology that replace the development of the economic and politic sector. During the recent four decades, globalization has been significantly influencing the world of integration of economy in all nations. Moreover, it is revealed that the economy of advanced and developed countries was improving with the ratio of the market on the gross domestic product (*Produk Domestik Bruto* or *PDB*) 50 percent in 1970 to 2009 (Mazumder, 2017). In response to globalization, every nation is responsible for designing a concept for the youth through a better education system. The governments must, however, support the educational development program since this is also the part of their responsibilities. However, the authority of a nation depends on the capital which Borneo (as cited in Mazumder, 2017) states. Further, Borneo (as cited in Mazumder, 2017) argues that the capacity of a nation is determined by the integration of individuals, finance, and resources. In other words, the role of a nation regarding the authority of policy is essential to prepare competent and professional individuals for the sustainable development of the nation. This also leads to a competitive nation.

5. Conclusion

Vocational education refers to a concept of integrating hands-on job training or practice in the regular class to prepare competent and professional individuals. The development of science and technology is considered as a solution to the issue around the development of education and globalization in the society which later brings challenges to the people. The challenges involve (1) conducting an extensive and in-depth research on the transformation of the social structures of a society, (2) improving the productivity of the nation for a sustainable development, (3) the emergence of new movements regarding science and technology, and (4) competition that becomes stiffer in the globalized world.

References

[1] Calhoun, C.C. and Finch, A.V. 1982. *Vocational Education: Concept and Operations* California: Wads Worth Publishing Company.

[2] Djojonegoro, Wardiman. 1998. *Pengembangan Sumber Daya Manusia melalui SMK* [Developing Human Resources through Vocational High School]. PT. Jayakarta Agung Offset. Jakarta

[3] Kurniawan, Khaerudin, "*Arah Pendidikan Nasional Memasuki Milenium Ketiga*" [Proceeding to the Third Millennium: The Progress of National Education], Suara Pembaharuan, Januari 1999.

[4] Mazumder, Sandeep: 2017 " *The Impact Of Globalization On Inflation In Developing Countries*, Journal Of Economic Development Vol.42, Number 3, Sept.2017

[5] Nena, Madlena, 2015: "Professional Insertion of Highly Qualified Human Resources. Study – Romania "Procedia - Social and Behavioral Sciences 191 (2015) 1891 – 1896

[6] Pasban, Mohammad, Sadegheh Hosseinzadeh Nojedeh, 2016 "A Review of the Role of Human Capital in the Organization: Procedia - Social and Behavioral Sciences 230 (2016) 249 – 253.

[7] Rahdiyanta, Dwi: 2014 Tantangan Pendidikan Teknologi Kejuruan dalam Era Global [Challenges of Technology Vocational Education in Globalization]: Prosiding APTEKINDO ISBN: 978-602-72004-0-1

[8] Saeed Aamir *et.al*: 2015 " *Impact of Globalization and the Role of International Agencies in Education Policy Making Process of South Asian countries – a Case of Pakistan*, A Research Journal of South Asian Studies **Vol. 30, No.2,** July – December 2015, pp.293 – 311.

[9] Tatiana, Blinovaa *et.al*, 2015 " *Vocational Education in the System of Determinants of Reducing Youth Unemployment: Interregional Comparisons*" Procedia - Social and Behavioral Sciences **214** (2015) 526 – 534

[10] Undang – Undang No. 20 Tahun 2003 "Tentang Sistem Pendidikan Nasional" [The Act Number 20 of 2003 Considering the National Education System].

[11] Yoosomboona, Sathaporn and Panita Wannapiroon, 2015 " *Development of a challenge based learning model via cloud technology and social media for enhancing information management skills*, Procedia - Social and Behavioral Sciences **174** (2015) 2102 – 2107.

Error analysis of mathematics students who are taught by using the book of mathematics learning strategy in solving pedagogical problems based on Polya's four-step approach

Budi Halomoan Siregar, Izwita Dewi, Ade Andriani

Department of Mathematics Education, Faculty of Mathematics and Natural Science, State University of Medan

budi_or_fuzzy@yahoo.com

Abstract. The purpose of this study is to analyse the types of students errors and causes of them in solving of pedagogic problems. The type of this research is qualitative descriptive, conducted on 34 students of mathematics education in academic year 2017 to 2018. The data in this study is obtained through interviews and tests. Furthermore, the data is then analyzed through three stages: 1) data reduction, 2) data description, and 3) conclusions. The data is reduced by organizing and classifying them in order to obtain meaningful information. After reducing, then the data presented in a simple form of narrative, graphics, and tables to illustrate clearly the errors of students. Based on the information then drawn a conclusion. The results of this study indicate that the students made various errors: 1) they made a mistake in answer what being asked at the problem, because they misunderstood the problem, 2) they fail to plan the learning process based on constructivism, due to lack of understanding of how to design the learning, 3) they determine an inappropriate learning tool, because they did not understand what kind of learning tool is relevant to use.

Keyword: error analysis, problem solving, polya four-step approach

1. Inroduction

Problem solving is an activity to overcome any problem by using concepts and rules, finally expected to get a solution to reach the goal (Dewi, et al 2017). Wilson (1993: 57) stated, "The problem solving has a special interest in learning mathematics. The main purpose of teaching and learning of mathematics is to develop the ability to solve complex mathematical problems". Furthermore, the approach of problem solving applied in this study is a Polya's four-step approach to problem solving.

Pedagogic ability is a competency related to learning or education. This ability includes an understanding of the character, the level of development of psychology, educational concepts, teaching methods which are in accordance with student development, etc. Finally, it can be concluded that pedagogic competence is the ability of a teacher to plan the learning and apply learning strategies which in accordance with the development and character of learners.

There are several reasons why teachers should master pedagogic competence. Firstly, the teacher will understand easily the difficulties faced by the students when they have problems in understanding the subject matter, able to stimulate the development of students well, able to apply active learning, able to apply various models of learning in accordance with the material and development of learners, able to build a conducive atmosphere when the learning process.

Pedagogics problems is related to education and how to educate. Furthermore, the ability to solve pedagogic problems is one of the most basic and important competencies to be mastered by teachers, in order to be able to educate them according to the rules, principles and norms.

On the other hand, as a researcher and lecturer in mathematics teaching and learning strategy of mathematics, has identified some student mistakes in solving pedagogic problems. These errors are known based on interviews and answer sheets on quiz questions. Based on these errors, it is necessary to analyze the types of errors and sources of cause. Then, the results of this analysis are used to improve the learning strategy, so that students do not make the same mistakes.

2. Theoretical Review

The type of this research is qualitative descriptive research in which aims to identify errors and causes in solving pedagogic problems. There are 34 students as subjects in this research, they have been learning about mathematics learning strategy in academic year 2017-2018. John C. Creswell (2009: 22) said that Qualitative research is a means for exploring and understanding the meaning of individuals or groups as a social or human problem. The process of research involves emerging questions and procedures, data typically collected in the participant's settings, data analysis inductively building from particulars to general themes, and the researcher making interpretations of the meaning of the data. The final written report has a flexible structure. Those who engage in this form of inquiry support a way of looking at research that honors an inductive style, a focus on the individual meaning, and the importance of rendering the complexity of a situation.

Data were obtained through a test and interviews. The test is given to obtain the kinds of errors in solving pedagogic problems. Before the tests were conducted, the questions were validated by three lecturers who were experts in the field of evaluation. Morever, interview is an exchange of information through a question and answer process conducted by two people to obtain meaningful information in a particular topic (Esterberg in Sugiyono: 2012, 317).

After obtaining the data, then analyzed through three stages, namely: 1) data reduction, 2) displaying and explaining data and 3) drawing conclusions. Furthermore, the determination of the student's answer score is done by applying Polya's four-step approach to problem-solving steps. Then, the test results are consulted on the scoring guidelines of problem solving test.

After reducing, then the data is presented in a simple form of narration, tables and diagrams. To determine the level of problem solving ability of each indicator then used the following formula: $NP_k = \frac{R_k}{SM_k} 100\%$ (Purwanto, 2009: 102). Where NP_k: The percent value to be determined on k indicator, k = 1, 2, 3, 4. R_k: score obtained by student on k-indicator, SM_k: maximum score on k-indicator. After the score is obtained, then consulted on 5 categories, they are very high, high, medium, low, and very low.

The third stage is the process of concluding the results of student error analysis. This stage is done by analyzing the results obtained based on previously presented data.

3. Result And Discussion

After the student answer sheets are checked and reduced, then the data obtained in table 1.

Table 1. Ability of understanding the problems, plan, carry out the plan, and looking back

Value Interval (%)	Types of category	Percentage of student achievement			
		Indicator 1	Indicator 2	Indicator 3	Indicator 4
90-100	Very High	24%	18%	18%	18%
80-89	High	35%	24%	24%	21%
65-79	Medium	32%	26%	26%	26%
55-64	Low	9%	32%	32%	35%
0-54	Very Low	0%	0%	0%	0%

3.1. Error in Preparation: Understand The Problem

Based on table 1, there are some students who make mistakes in understanding of pedagogic problems with different error levels. 32% and 3% of students who have not been able to understand the problem well, in which concist of in the medium and low category, respectively. To find out the types and causes of the error, student answer 5 (M5) is shown as below.

a. Understanding of pedagogic problems Based on the problem can be known: • The material already taught is the concept of cylinder volume • The following material to be taught is the concept of cone volume • It is recommended to use learning tools and learning models • The learning process to be implemented is based on constructivism. The question is: What is the relation of contructivism theory to the concept of cylinder and cone volume.

According to the answer sheet, it is known that M5 has been able to identify what are known on the problem. However, M5 has not been able to identify what are being asked on the problem well. To find out the error, the results of interviews between researchers and M5 are shown below.

Researcher: According to your answers, you have been able to identify information on the problem. However, your explanation is quite short at the point of what is being asked on the problem. Why?
Student 5: Yes sir, I think that explanation already represents what is asked on the problem.
Researcher: Are you sure of your answer?
Student 5: I am sure that my answer is correct
Researcher: Please re-read the problem, I give you time for 7 minutes.
Student: ok sir
Researcher: After reading it, how do you think?
Student: My answer is correct sir, but there is one thing that must be added.
Researcher: What is it?
Student: The problem instructs to describe the initial, core, and closing activities.
Researcher: Why did not you write down this point at that time?
Student: I did not identify this point at that time sir. I was not careful to understand the problem.
Researcher: According to your answer sheet, you wrote in the point of what beeing asket "The relationship between the constructivism theory and the material of the cylinder and cone. Are you sure of your answer?
Student 5. I'm sure sir. Etc....

According to interview, it can be seen that the student misunderstood at the point of what being asked on the problem. Furthermore, after the researcher asked to re-read the problem, the student was aware of his mistake and able to fix it by himself. This is because the student has not been careful to understand the problem during the test.

After analyzing all student answer sheets, it can be concluded that there is a uniform error. They make mistakes in identifying and understanding what is being asked on the problem, because they do not read the problem carefully.

3.2. Error in Thinking Time: devising a plan

According to table 1, it can be concluded that there are 26% and 11% of students in medium and low category in devising a plan. This data shows that there are still some students who are wrong in making plans. In order to know the types of errors, the following answer sheet shown below.

a. Devising a plan • Design a student worksheet based on constructivism. • Using the PBL Model • Develop a plan of learning based on PBL syntax • Students will be grouped, 3-4 students per group.

IOP Conf. Series: Journal of Physics: Conf. Series **970** (2018) 012004 doi:10.1088/1742-6596/970/1/012004

- Prepare cylinder and cone objects where both types of objects have the same base area and height.
- In the learning process, the teacher will show cones and cylinders, then explain the formulas of each object.
- Teacher will facilitate students in experimenting.

The results of interview between lecturer and student 9 shown as follows.
Researcher: After checking your answer sheet, the result is good at point b.
Student: Thank you sir.
Researcher: However, do you only need cylinders and cones in experimenting?
Student: You are right Sir, I just need those two learning tools
Researcher: How the use of cylinders and cones in the learning process that you will do.
Student: I will show the cylinders and cones in front of the class. It is expected that they will recognize the two objects well.
Researcher: In addition, what are the cylinders and cones for? And what do you mean by the answer at the last point?
Student: I will give cylinder and cone to each group. Then I will ask them to disassemble these objects, so they will know the shape of skin both of objects.
Researcher: Then when do you plan to teach the concept of cone volume?
Student: After recognizing the shape of cylinder and cone well, then I will explain the concept of cone volume, and then give some problems related to it. Etc....

Based on the answer sheet and the results of the interview, it can be seen that the learning tool used in the learning process is incomplete. This is due to that the student does not understand how to use the learning in finding the concept of cone volume.

After analyzing the answer sheets and interview, there are several types of errors found as shown in Table 2.

Table 2. Types and causes of errors in devising plans.

Types of errors	Causes of the errors
They make mistake to plan a constructivism-based learning	They did not understand how to design a constructivism-based learning
They determine an inappropriate learning tool	Students did not know well what learning tool is appropriate to use
They demonstrate improperly the use of learning tools	They did not understand how to use the tool in finding the concept of cone volume

3.3. Error in Insight: carrying out the plan

Table 1 shows that there are 32% and 26% of students in low and medium categories in carrying out problem solving. This data shows that there are still many students who make mistakes in that step. To find out the type of errors and its causes, the following answer sheet and interviews are shown.

b. Carry out the plan
Initial activity
Distribute student worksheets and learning tools, then explain the competencies to be achieved.
Core activities
The teacher will explain material: "if there are any cylinders and cones with the same base and height, then: cylinder volume = 3xthe volume of the cone, so it can be concluded that the $cone\ volume = \frac{1}{3} volume\ of\ the\ cylinder = \frac{1}{3}\pi^2 t$"
After the students understand the concept, students are then instructed to do the following experiment: Fill mung beans into the cone until full. Then pour the contents of the cone into the cylinder. This activity is repeated until the volume of the cylinder is full. Based on the experiment, students are expected to understand that the cone volume is equal to $\frac{1}{3}$ the volume of the cylinder. So

> the formula is obtained: $cone\ volume = \frac{1}{3} volume\ of\ the\ cylinder = \frac{1}{3}\pi^2 t$"
> Closing Activity
> The teacher and the students will conclud the results of the experiment.

Here are the results of interviews between lecturer and student 21.
Researcher: According to your answer sheet, you plan that the teacher will explain the material before the experiment. Why?
Student 21. I think if i do it, the experiments process can be run smoothly.
Researcher: May you explain what are the principles of constructivism-based learning?
Students: In essence: knowledge must be constructed by the students themselves
Researchers: do you believe by explaining the material at the beginning of the activity can stimulate the students to construct the knowledge by themselves?
Student: ohh i am sorry Sir. I just realized that my answer is wrong.
Researcher: Why do you say like that? What is wrong exactly?
Student: teachers should stimulate students to construct their own knowledge. The experiment should be done at the beginning of the activity. With such that, it is expected that they will find the concept of cone volume. Etc....

Based on the answer sheet and interview, it can be seen that the design of core activities does not conform to the principles of constructivism-based learning. It is recommended that the students should conduct an experiment in the beginning activity, so that they will be able to construct the knowledge by themselves.

After analyzing the data, the types of errors and causes are shown below.

Table 3. The types of errors and causes in carrying out the plan

Type of error	Cause of the error
Not implementing constructivist-based learning	They did not understand well how to design the learning proscess based constructivism
The use of learning tools is irrelevant to find the concept	They did not understand well how to use the learning tools to find the concept of cone
They make mistake in designing of core activity.	Not using a relevant model, so the learning proscess is not organized according to the principles of constructivism-based learning

3.4. Error in verification: looking back

Table 1 shows that 35% of students still make a lot of mistakes in verification proscess. Here is the answer sheet of student 30 (M30).

> d. look back.
> Initial activities, core activities, and the closing have been described at the planning step. These three activities have been designed to be interrelated and structured based on constructivism-based learning.

The interview between researcher with student 30 shown below.
Researcher: Why do you check on the beginning, core, and closing activities only?
Student: I think by descibing these three sections can already represent the answer point.
Researcher: Do you think by the explanation that means you have already checked your answer?
Student: I doubt sir.....etc
Researchers: Do you need to ensure that the learning process has applied the constructivism based-learning?
Student: Yes i do, because this point asked on the question.
Researcher: What else is needed to be explained in verification?
Student: I think it is important to check whether to find the cone formula out has been associated with the concept of cylinder volume. Etc....

After analyzing the data, the following types of errors and causes are shown below.

Table 4. The types of errors and causes in looking back

Type of error	Cause of the error
The answer is still very common, so it is still not check the main point of the problem.	Students do not understand the technique and procedures to look back the answers
The student's answer in the step c is not synchronized with point d (verification)	Students can not be able to synchronize the answer point d with the answer point c.
They did not check their answer completely	They did not verify the all answer

4. Conclusion

Based on result and discussion, the type of student errors in solving pedagogic problem based on Polya procedure were: 1) the student made a mistake to answer what being asked at the problem, this is because they misunderstood at the problem, 2) they made a mistake to plan the learning proscess based on constructivism, due to lack of understanding of how to design the learning, 3) they chose an inappropriate learning tool, because they did not know what kind of learning tool is relevant to use, 4) They were wrong in sorting and organizing the learning process, because they did not yet understand how to use the learning model appropriately.

Based on the above conclusions, the researchers recommend to: 1) to improve the students' ability to design the active learning, 2) to increase students' ability to determine the appropriate learning tools, 3) improving students' ability to apply a learning model.

References

[1] Arikunto, S.2010. *Prosedur Penelitian Suatu Pendekatan*. (Yogyakarta: RinekaCipta). P317

[2] Ario M. 2016. *Analisis kemampuan penalaran matematis siswa SMK setelah mengikuti pembelajaran berbasis masalah*. Jurnal Ilmiah Edu Research Vol.5 No.2. Dec 2016 P125)

[3] Creswell, John C. 2009. *Research design: Qualitative, quantitative, and mixed methods approaches, 3rd ed.* (Los Angeles: SAGE Publications, Inc.)

[4] Dewi I, Andriani A, Halomoan B. 2017. *Strategi Belajar Matematika Berbasis HOTS*: (Medan).

[5] Polya G. 1973. *How To Solve it: A New Aspect of Mathematical Method.* (New Jersey, USA: Pricenton University Press).

[6] Purwanto, (2009), *Evaluasi Pengajaran*. (Bandung: Rosdakarya). P.102

[7] Suherman E. 2003. *Strategi Pembelajaran Matematika Kontemporer.* (Bandung: Jurusan Pendidikan Matematika UPI).

[8] Wilson. 1993. *Mathematical Problem Solving*. (New York: Macmilan Publishing Company).

Moh's hardness Scale and micro vicker's hardness study of bgo and lyso inorganic scintillators

Wuttichai Chaiphaksa[1,2*], Pruittipol Limkitjaroenporn[1,2], Hong Joo Kim[3] and Jakrapong Kaewkhao[1,2]
[1]Physics Program, Faculty of Science and Technology, Nakhon Pathom Rajabhat University, Nakhon Pathom 73000, Thailand
[2]Center of Excellence in Glass Technology and Materials Science (CEGM), Nakhon Pathom Rajabhat University, Nakhon Pathom 73000, Thailand
[3]Department of Physics, Kyungpook National University, Daegu 702-701, Korea
wuttichai.chaiphaksa@gmail.com

Abstract - BGO and LYSO inorganic scintillation crystals were chosen in this paper for density and hardness properties investigation. Moh's hardness scale was determined by scratching test which of the standard minerals. The micro vicker's hardness was measured with a diamond indenter and the test load was used at 0.1 kgf. The results show that, the moh's hardness scale of LYSO crystal are higher than BGO crystal which correspond with micro vicker's hardness values. The density of scintillation crystals were measured and found to be 7.1 g/cm^3 for BGO crystal and 7.15 g/cm^3 for LYSO crystal.

1. Introduction

Inorganic scintillator are widely used in many application for engineering, high energy physics, oil exploration and nuclear medicine, i.e., positron emission tomography (PET) and single photon emission computed tomography (SPECT) [1-2]. Bismuth germinate ($Bi_4Ge_3O_{12}$) and lutetium yttrium oxyorthosilicate (LYSO) are a good choice used in commercial PET technology [3-4]. The choice of the detector is based on several characteristics such as scintillation decay time, light output, energy resolution of the detector, stopping power and also including hardness test [5]. The hardness test is a mechanical test for material properties which are used in engineering design, analysis of structures and materials development. The principal purpose of the hardness test is to determine the suitability of a material for a given application and including with the relationship between hardness and other properties of material [5]. In this paper present the micro vicker's hardness test and moh's hardness scale of BGO and LYSO inorganic scintillation crystals. Moh's hardness scale was determined by scratching test which of the standard minerals. The micro vicker's hardness was measured with a diamond indenter and the test load was used at 0.1 kgf. The density of scintillation crystals were also measured.

2. Experiment

The BGO and LYSO crystals with size of 10×10 cm^2 and $10\times10\times10$ cm^3 respectively, were used in this experiment. The micro vicker's hardness test method consists of indenting the test material with a diamond indenter, in the form of a right pyramid with a square base and an angle of 136 degrees between opposite faces subjected to a load not exceeding 1 kgf. The full load is normally applied for 10 to 15 seconds. The two diagonals of the indentation left in the surface of the material after removal of the load are measured using a microscope and their average calculated. The area of the sloping surface of the indentation is calculated. The Vickers hardness is the quotient obtained by dividing the kgf load by the square mm area of indentation follow as [5]:

$$HV = \frac{2F\sin\frac{136^\circ}{2}}{d^2} \quad (1)$$

$$HV = 1.854\frac{F}{d^2} \quad (2)$$

where F is load in kgf, d is arithmetic mean of the two diagonals, d1 and d2 in mm, HV is micro vicker's hardness. The Moh's hardness scale, it simply consists of 10 minerals arranged in order from 1 to 10. Diamond is rated as the hardest and is indexed as 10 and talc as the softest with index number 1. The hardness is determined by finding which of the standard minerals the test material will scratch or not scratch. The hardness will lie between two points on the scale, the first point being the mineral which is scratched and the next point being the mineral which is not scratched. In this study, focus on density and hardness properties investigation of BGO and LYSO crystals. The density was measured by Archimedes principle, the weights of the prepared samples were measured in air and in mineral oil (ρ_b = 0.825 g/cm3) using a 4-digit sensitive microbalance (AND, HR200). The micro vicker's hardness test method using digital display micro hardness tester (HVS-1000) was measure 6 side per sample and from each side were measured 5 point per side then average. The Moh's hardness scale was used 10 minerals arranged in order from 1 to 10 for measurement.

3. Results and discussion

Density and hardness properties

The density of BGO and LYSO crystals by Archimedes principle are shown in Figure 1. It was found that LYSO crystal is slightly higher value than BGO crystal. The value corresponds to 7.15 g/cm^3 for LYSO crystal and 7.1 g/cm^3 for BGO crystal.

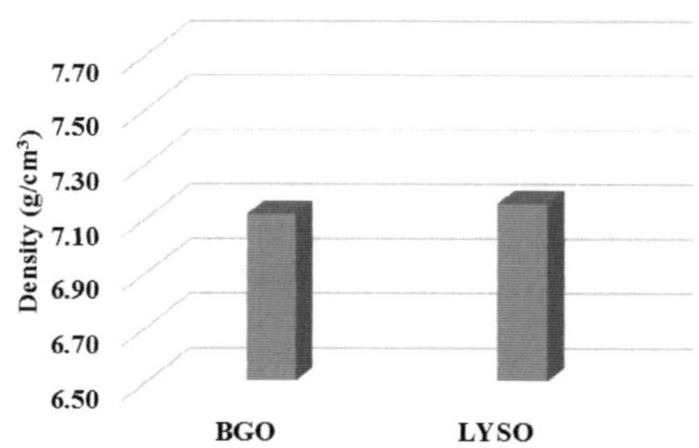

Figure 1: Density of BGO and LYSO crystals

Figure 2 shows the micro vicker's hardness of BGO and LYSO crystals, measured using digital display micro hardness tester and the test load was used at 0.1 kgf. It is seen that the micro vicker's hardness of LYSO crystal are higher than BGO crystal. Figure 3 shows that the moh's hardness scale of BGO and LYSO crystals, measured by the standard minerals. It was found that the moh's hardness scale of LYSO crystal are higher than BGO crystal which correspond with micro vicker's hardness values.

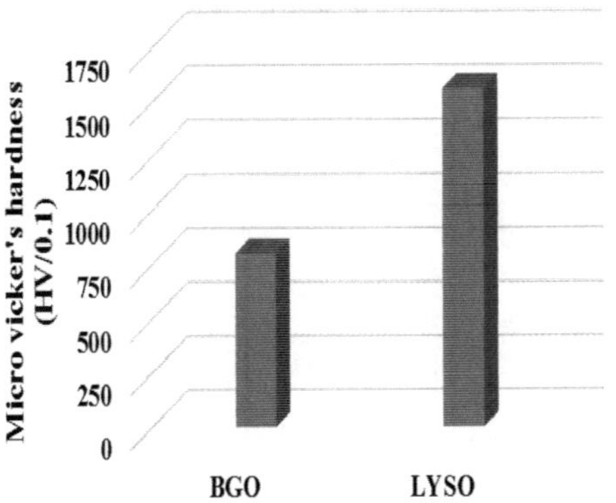

Figure 2: Micro vicker's hardness of BGO and LYSO crystals

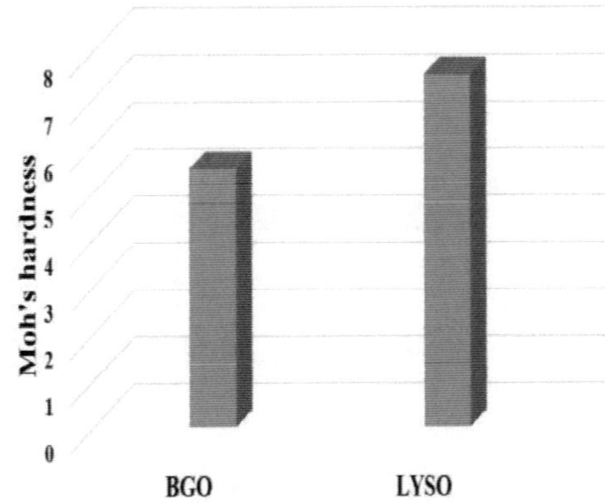

Figure 3: Moh's hardness scale of BGO and LYSO crystals

4. Conclusion

The density and hardness properties of BGO and LYSO inorganic scintillation crystals have been studied. Moh's hardness scale was determined by scratching test which of the standard minerals. The micro vicker's hardness was measured with a diamond indenter and the test load was used at 0.1 kgf. The results show that, the moh's hardness scale of LYSO crystal are higher than BGO crystal which correspond with micro vicker's hardness values. The density of scintillation crystals were measured and found to be 7.1 g/cm^3 for BGO crystal and 7.15 g/cm^3 for LYSO crystal.

Acknowledgments

The authors gratefully acknowledge the National Research Council of Thailand (NRCT) for supporting this research and Nakhon Pathom Rajabhat University (NPRU) for facilities.

References

[1] Van Eijk, C.W.E., "New inorganic scintillators aspects of energy resolution," *Nuclear Instruments and Methods in Physics Research A*, vol. **471**, pp. 244-248, 2001.

[2] Van Eijk, C.W.E., "Inorganic-scintillator development," *Nuclear Instruments and Methods in Physics Research A*, vol. **460**, pp. 1-14, 2001.

[3] Gopal, B. S., *Basics of PET Imaging: Physics*, Chemistry and regulations, 2nd ed., Springer, pp. 18-20, 2010.

[4] M. Moszynski, "Energy resolution and non-proportionality of scintillation detectors new observations, *Radiation Measurements*," vol. **45,** pp. 372-376, 2010.

[5] M. J. Weber, "Inorganic scintillators: today and tomorrow," *Journal of Luminescence*, vol. **100** pp. 35-45, 2002.

[6] J. Newby, *Metals Handbook*, Mechanical Testing, 9th ed., Vol. **8**, 1990.

IOP Conf. Series: Journal of Physics: Conf. Series **970** (2018) 012006 doi:10.1088/1742-6596/970/1/012006

Chitosan/graphene oxide biocomposite film from pencil rod

S Gea,[1*], J N Sari[1], R Bulan[1], A Piliang[2], S A Amaturrahim[1], Y A Hutapea[1]

[1]Department of Chemistry, Faculty of Mathematics and Natural Sciences, University of Sumatera Utara, Jl. Bioteknologi No. 1 Medan, 20155, Indonesia.
[2]Department of Physics, Faculty of Mathematics and Natural Sciences, University of Sumatera Utara, Jl. Bioteknologi No. 1 Medan, 20155, Indonesia.

*Correspondence email: s.gea@usu.ac.id

Abstract. Graphene Oxide (GO) has been succesfully synthesized using Hummber method from graphite powder of pencil rod. The excellent solubility of graphene oxide (GO)in water imparts its feasibilty as new filler for reinforcement hydrophilic biopolymers. In this research, the biocomposite film was fabricated from chitosan/ graphene oxide. The characteristics of graphene oxide were investigated using Fourier Transform Infrared (FT-IR) and X-ray Diffraction (XRD). The results of the XRD showed graphene structur in 2θ, appeared at9.0715°with interlayer spacing was about 9.74063Å. Preparation films with several variations of chitosan/ graphene oxide was done by casting method and characterized by mechanical and morphological analysis. The mechanical properties of the tensile test in the film show that the film CS/ GO (85: 15)% has the optimum Young's modulus size of 2.9 GPa compared to other variations of CS / GO film. Morphological analysis film CS/GO (85:15)% byScanning Electron Microscopy (SEM), the obtained biocomposites film showed fine dispersion of GO in the CS matrix and could mix each other homogeneously.

1. Introduction

Graphene, as an unique material, provides several interesting characteristics because of its structure which has hexagonal structure forming planar sheet of all-sp^2as size as an atom [1]. The most interesting characteristic, the mechanical strength of graphene is higher than steel due to its structure indicating this material can be used as reinforcement in fabricating composite. However, in utilizing it as reinforcement in biocompatible composite, it is more suitable to combine the graphene oxide (GO) than graphene due to the higher presence of oxygen functional groups particularly hydroxyls and epoxides. Chemically speaking, the existence of these two characteristics – either the structure and functional groups --, provides higher possibility of intercalation of various molecules so it can be functioned in fabricating composite. Therefore, in this paper we present the utilization of graphene oxide as filler to enhance the mechanical strength of chitosan.

Synthesizing graphene has invited many scientists. Graphene is considered as a material which has unique properties either in mechanical and electrical. It has two-dimensional structure with sp2 hexagonally packed in density and it was predicted in 1940 from graphite theoretically[2]. Sixty years after predicting, single graphene sheets had been prepared by Novoselov and Geim of the University of Manchester[3]. Many studies have reported that graphene's properties for either in academic or industrial field. It is potentially to be designed for supercapacitor[4]; coated in composite with conductive properties[5]; and as catalyst[6].

Graphene is a monolayer graphitic carbon which is formed inside the graphite. This deposition then can be separated by performing the two steps which are Brodie oxidation and Hummers method. In synthesizing the graphene, the preparation of graphene oxide must be conducted in form of dispersions in water and other solvents[7]. However, the graphene oxide (GO) membranes and the precursor of graphite oxide generated significance differences in hydration and solvation properties [8]. Therefore, intercalation process can be occurred exactly the same either in water or alcohol condition depending on the numbers of monolayers which are corresponded, but the solvents' volume are varied as the temperature and pressure conditions [9].

The preparation of graphene oxide (GO) widely used is exfoliation by performing Hummers method. The reaction involves the intercalation process which is essential in obtaining the quantities of graphene oxide. Subsequently, the preparation must be conducted in electrochemical reaction so the reduction from graphite to graphite oxide is proceeded properly[10]. Thus, the graphene oxide can be derived from graphite oxide by instrument teststhroughout the presence of functional groups.

Chitosan is a derivative of chitin compounds which is produced by de-acetylation involving high concentrate of base solution. This material is classified as organic material as the chemical composition is composed of linear polysaccharide with glucosamine[11]. The linearity of chemical structure produces interesting properties, such as good flocculants and chelating properties. While in the physical form, chitosan is hard and commonly found as skeleton of invertebrate, such as *Arthropod Sp*, but the most resources of chitosan can be found in *Crustacean Sp*. Surprisingly, chitosan comprises two hydroxyl group functions and one free amino group indicating the material is occasionally dissolved in neutral pH of water[12]. And more importantly, chitosan is an organic material with reactive functional groups particularly in the amino compounds so it is non-toxic or biocompatible.

The preparation of chitosan and graphene oxide are respectively prepared from Blankasand pencil's graphite. The fabrication of composite is performed by using casting method with certain percentage of compositions. The bio-composite that has been prepared therefore is investigated to understand the morphological properties with Scanning Electron Microscope (SEM). In addition to understand the mechanical properties, we performed mechanical test including tensile test and analyze the data with software OriginPro 2016 (32-bit) b9.3.226.

2. Materials and methods

2.1. Materials and Instruments

AnalyticalBalance, Oven (Carbolite), Mechanical Stirrer, Universal pH, A thermometer, A hot plate, magnetic Stirrer, Ice bath, Cuvette, Bottles of distilled water, A spatula, Sentrifugator VS-GOOCFi, Ultrasonication instruments, a set of SEM JSM- 35C Shumandju, Equipment tensile Gotech AL 7000 M, a set of XRD Shimadzu XRD-6100, a set of FTIR Shimadzu IR prestige-21, chitosan solid powder,graphite of pencil rod,Distilled water (H_2O), H_2SO_4 98%, $KMnO_4$, $NaNO_3$,H_2O_2 30%, CH_3COOH 2%.

2.2. Preparation of grapehene oxide

As many as 2 grams of graphite were putted in beaker glass and then those were added with 1 gram of Sodium Nitrate ($NaNO_3$),then the mixture powder was poured with 46 ml of sulphate acid (H_2SO_4) 98%. During the reaction, the solution was in exotermic condition so it must be stirred in the ice-bath to maintain temperature while the $KMnO_4$ was added after the temperature being controlled by the ice bath to 20°C. The solution then was moved from the ice-bath and it was heated on temperature of 35°C for twenty hours. Afterward, the solution then was poured with 92 ml of distilled water gradually while at the same time it was stirred for 15 minute to achieve neutral conditions. Hence, the solution was added with hydrogen peroxide (H_2O_2 30%) before being centrifuged by centrifuge with angular velocity for 7200 rpm for 10 minutes. After being centrifuged the solution was comprised by solution and precipitate. Then, the precipitate was washed with warm distillate water to increase the pH near 7.

The obtained yellow- brown powder was re-dispersed into distillate water while at the same time stirred until homogeneus, followed by treatment usingmild ultrasound for 30 min. Graphene oxide powder was obtained after dried at 105°C of the suspension.

2.3. Preparation of chitosan/graphene oxide biocomposite film

As many as 2 grams of chitosan were dissolved into 200 ml of aqueous acetic acid 2% to prepare chitosan solution. The mixture was stirred continuously at room temperature. The graphene oxide powder (0.2 g) was dispersed into 20 ml of distilled water. Thus, the grahene oxide was added in chitosan solution under stirring at room temperature for 1 hour followed by sonication for 1 hour with 50/60 Hz to ensure a homogeneous dispersion of chitosan/graphene oxide in solution. The mixed solution was cast on glass plate and then dried in oven at a temperature 50°C for 8 hour. The films containing 0, 5, 10, 15, 20, 25 % GO were removed from the glass plate and cut.

2.4. Characterization

The Fourier Transform Infrared (FT-IR) spectra for graphite of pencil rod, graphene oxide, chitosan, and chitosan/graphene oxide biocomposite film were recorded on a Shimadzu IR prestige 21. FT-IR spectra were measured in the range 4000-500 cm^{-1}. Powder samples were mixed with KBr and pressed it into a thin pellet used for analysis.

To clarify the crystal structure of fine grahite,graphite of pencil rode, graphene oxide, wide-angle X-ray Diffraction (XRD) measurements were recorded using Shimadzu XRD-6100. The Cu-Kα was used as the source (wavelenght $\lambda = 0.1542$). the recorded region of 2θ was 5-45°.

The tensile properties of cast sample were measured using a Gotech Al 7000 M at room temperature. A 1 kN load cell was used and the strain rate was 5.0 mm/min. To ensure data accuracy and repeatily, as least 3 measurements were carried out for each biocomposite film.

The morphology of biocomposite films was observed using Scanning Electron Microscopy (SEM), JSM-35 C Shumandju. The films were frozen in liquid nitrogen, and immediately snapped and then sputtered with gold before observation.

3. Result and discusscion

Chitosan and graphene oxide can be well mixed and form homogeneously aqueous solution, and the obtained mixture is stable at the room temperature. After mixing, the solutions of chitosan or chitosan/ graphene oxide were casted onto substrates, there formed the films. FTIR studies confirmed the successful oxidation from graphite extracted from pencil rodto graphite oxide. The FT-IR spectrum of graphite of pencil rod which is displayed in figure 1 depicts astrong–OH peak at 3410.15 cm^{-1} and C=O group at 1620.1 cm^{-1} while the spectrum also shows a C-C peak at 1056.99 cm^{-1}. The FT-IR spectrum of graphene oxide depicts a strong–OH peak at 3394cm^{-1} and the other functional groups of C–O including COOH (1720 cm^{-1}), C–OH (1388 cm^{-1}), and C–O–C (1087 cm^{-1}) which are clearly visible. The spectrum also shows a C=C aromatic peak at 1580 cm^{-1} while,for chitosan, the peak at 3425.58 cm^{-1}which both of them are corresponding to N–H stretching vibration of amino group. The vibrations related with N-H bending and N-H deformation are at 1566 cm^{-1} and 1404.18 cm^{-1}, repectively. According to the FTIR's result, the peak of 1404.18 cm-1 indicates the presenceof vibration mode C-OH.The peaks characteristics absorption in the curve of CS/graphene oxide are approximately similar with the curve of fine chitosan as shown in Fig. 1. The FT-IR spectrum of the CS/graphene oxide compound shows a combination of characteristics which is similar to that of fine chitosan and graphene oxide which includes the broad absorption ofband is located at 3425.58 cm^{-1}, assigned to the mixture of the amine stretch from the chitosan to the OH groups in graphene oxide. The peak at 1620.21 cm^{-1} illustrates the presence of the COOH groups from graphene oxide and is downshifted due to hydrogenbonding between the graphene oxide and hexatomic ring of the chitosan.

On the other hand, the peak at 1566.2 cm^{-1}presents the C=C groups from graphene oxide and it also indicates the presence of N-H bonding that is resulted from chitosan. The spectrum of the N-H deformation from chitosan represented at 1411.89 cm^{-1}. The peak in the region about 1080.14 cm^{-1} is the indication of C–O–C stretching from the graphene oxide layers. The FT-IR results indicate the existence of the interaction between chitosan and graphene oxide.

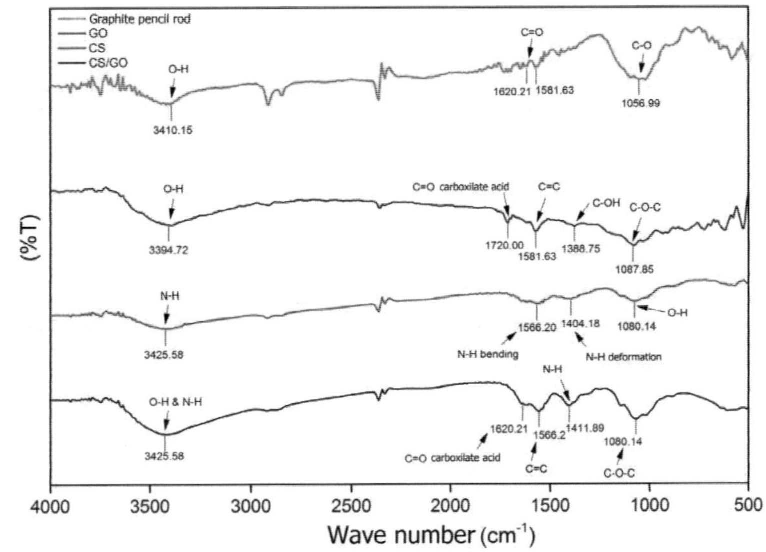

Figure 1. FT-IR spectrum from graphite of pencil rod, chitosan, graphene oxide, and chitosan/graphene oxide film

The XRD patterns of graphite of pencil rod, graphite powder, and graphene oxide are depicted in figure 2 performed in the range of 5°-70°. The diffraction angles of the graphite of pencil rod forms peak 26.369° with d-spacing 3.7721 Å. On the other side, the commercial graphite forms peak 26.6025° with d-spacing 3.34810 Å. That isproves that the graphite of pencil rod compound have a lot of percentage of graphite. Graphene oxide form a new peak is 9.0715° with d-spacing 9.74063Å. The change of angle XRD patterns are shifted toward the left shows the occurrence of a material change to the nature of crystallinityGraphene oxide is reduced (Taufantri, *et al.* 2016). Widening the distance between the layers of graphite pencils of 3.34810 Å into Graphene oxide at 9.74063Å. It occurs due to the formation phenol group, an epoxy group, ketone group, carboxyl group, and a carbonyl group. The addition of Hmolecules$O_2$and clustersof oxygen also causes graphene oxide has a d-spacing wider. (Taufanri, *et* al,2016).

IOP Conf. Series: Journal of Physics: Conf. Series **970** (2018) 012006 doi:10.1088/1742-6596/970/1/012006

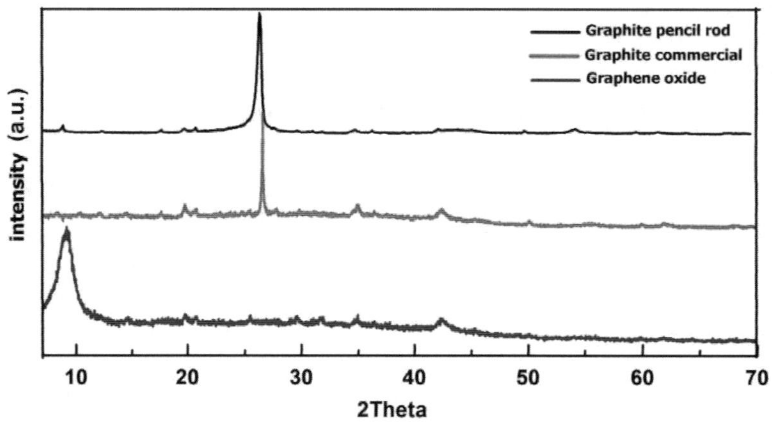

Figure 2. XRD of pencil rod graphite, graphite and graphene oxide

In this research, we performed tensile test to understand the mechanical strength of chitosan film and its variation based on the composition of the graphene oxide as the filler. The tensile test performed 2000 Kgf of force with speed approximately 5 mm for every minutes in condition of room temperature. The average film thickness of the film CS and CS / GO is 0.1 to 0.11 mm. The tensile strength of the film and the film biocomposite CS CS / GO can be seen in Table 1.

Table 1. Data Testing Mechanical properties of Graphene Oxide Film of Chitosan and CS/GO

No.	Variations of biocomposite films	Maximum strain (MPa)	Modulus *Young's* (GPa)
1	CS	5.83	1.0
2	CS/ GO 95: 5%	7,75	0 , 9
3	CS/ GO 90: 10%	10	2.0
4	CS/ GO 85: 15%	10.5	2.9
5	CS/ GO 80: 20%	5.68	0.9
6	CS/ GO 75: 25%	4.56	1.6

Table1 shows the addition of fillers on biocomposite ranging from weight variation of 5% increases the tensile strength of biocomposites CS/ GO ranging from 5.83MPa to 7.75MPa. The maximum tensile strength occurs in variations CS/ GO 85: 15% is equal to 10.5MPa while the minimum properties shown in variations of 75:25%. However, the tensile strength decreased with the addition of Graphene oxide by 20% and so on.The exact properties are also shown on the graph the stress-strain of chitosan films and films biocomposite CS/ GO displayed onFig.3.The Modulus *Young's* of pure chitosan equals to 1.0 GPa. With the addition of fillers that graphene oxide film inbiocomposite CS/ GO increases the Modulus Young drastically in the variation of CS/ GO 85: 15%, around 2.9 GPa.

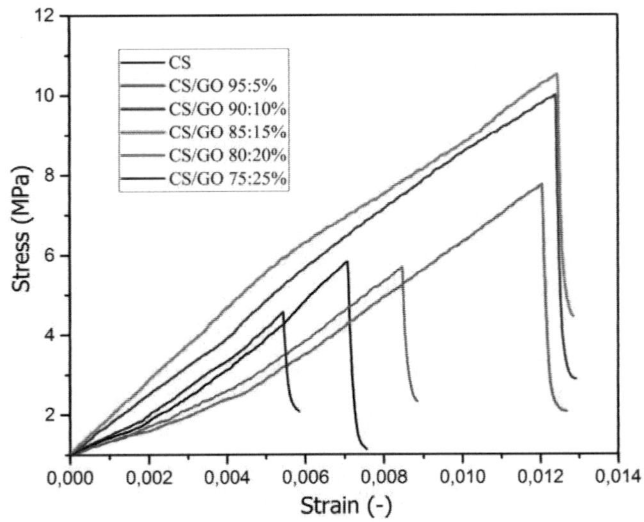

Figure 3. Graph Stress-Strain of Film and Film Chitosan Chitosan / Graphene Oxide

According to the Fig.3, the presence of graphene oxide as the reinforcement increases the stress-strain. But more importantly, the quantities of GO within the biocomposites are associated to the mechanical properties, in particular the tensile strength. The composition of CS/ GO with variation of 85/15% affects the tensile properties in maximum numbers. The increase in tensile strength of the resulting biocomposite films may be due to the strength of hidrogen chain attached to the chitosan and graphene oxide and homogeneous distribution of chitosan and a high compatibility between graphene oxide layer and matrix assisted by a high interfacial surface area. However, the increase in weight gain by 20% Graphene oxide effects tensile test for impairment and Young's modulus. This may be caused by the large value of viscosity.

The results of SEM analysis of the film provide direct information about the interaction interface of the film that can be seen in Figure 8, showing the surface of the film CS and CS/GO. In figure 4a, it can be seen the film's surface is clear without the presence of Graphene Oxide, while it also appears that the film does not have pores implying of high density caused by the dissolved in acetic acid 2%. In figure 4b. the surface of biocomposite CS/ GO with variation ratio of 85: 15% is illustrated mixedhomogenously implying the uniformity and homogeneity in the distribution of Graphene oxide on the film.And the cross-linking and physical chain chitosan structure in the CS/ GO (85:15) % is denser than the film CS, which predicts an increasing in the mechanical strength.

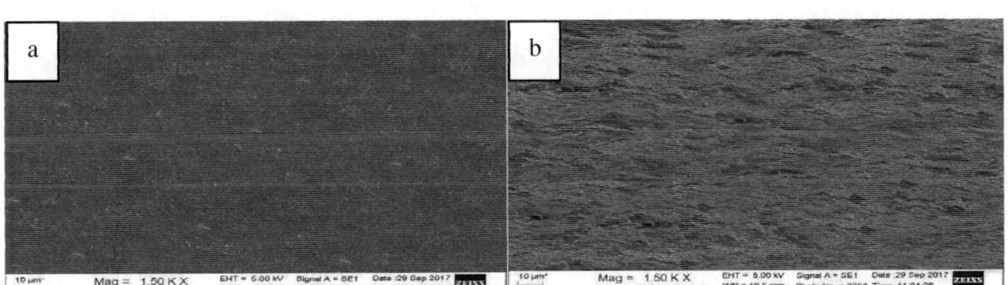

Figure 4. (a) SEM on the film CS, and (b) SEM CS / GO 85: 15%

4. Conclusion
The graphene oxide was successfully synthesized from pencils rod by using Hummer method, however, the quantities of the graphene oxide that has been extracted is small. Comparing from previous research, the graphene oxide synthesized from pencils rod is smaller than those which from natural graphite significantly. Furthermore, the crystallinity of both graphite and graphene synthesized from pencils rod are similar to those which are synthesized from natural graphite. The mechanical properties of film biocomposite is influenced by the presence of graphene oxide as reinforcement. Based on the tensile test, the optimum composition of chitosan and graphene oxide is 85:15%, which contributed for 10.5 Mpa and 2.9 GPa of Modulus Young. While the minimum numbers of tensile test and modulus Young is contributed by the composition of 75:25%, accounted for 4.56 and 1.6, respectively.

5. Acknowledgement
The authors would like to address a gratitude to the head of physical chemistry laboratory for facilitating the research

6. References

[1] L. He, H. Wang, G. Xia, J. Sun and R. Song, "Chitosan/graphene oxide nanocomposite films with enhanced interfacial interaction and their electrochemical applications," *Applied Surface Science,* no. 314, pp. 510-515, 2014.

[2] P. R. Wallace, "The Band Theory of Graphite," *American Physical Society,* 1946.

[3] K. S. Novoselov, A. K. Geim, S. V. Morozov, D. Jiang, Y. Zhang, S. V. Dubonos, I. V. Grigorieva and A. A. Firsov, "Electric Field Effect in Atomically Thin Carbon Films," *Science,* vol. 306, no. 5696, p. 5, 2004.

[4] D. Li, C. Yu, M. Wang, Y. Zhang and C. Pan, "Synthesis of nitrogen doped graphene from graphene oxide within an ammonia flame for high performance supercapacitors," *RSC Advances,* no. 98, pp. 55394-55399, 2014.

[5] X. Wang, W. Xing, L. Song, H. Yang, Y. Hu and H. G. Yeoh, "Fabrication and characterisation of graphene-reinforced waterborne polyurethane nanocomposite coatings by the sol-gel method," *Surface and Coatings Technology,* vol. 206, no. 23, pp. 4778-4784, 2012.

[6] S. Basu, M. C. Lee and Y.-H. Wang, "Graphene-based electrodes for enhanced organic thin film transistors based on pentacene," *Physical Chemistry Chemical Physics,* no. 31, pp. 16701-16710, 2014.

[7] P. H. Boehm, A. Clauss and U. Hofmann, "Graphite Oxide and Its Membrane Properties," *J. Chim. Phys,* vol. 58, no. 1, p. 7, 1961.

[8] A. V. Talyzin, G. Mercier, A. Klechikov, M. Hedenstrom, D. Johnels, D. Wei, D. Cotton, A. Opitz and E. Moons, "Brodie vs Hummers Graphite Oxides for Preparation of Multi-Layered Materials," *Carbon,* vol. 115, pp. 430 - 440, 2017.

[9] A. V. Talyzin, M. S. Luzan, T. Szabo, D. Chernyshev and V. Dimitriev, "Temperature Dependent Structural Heating of Hydrated Graphite Oxide in H20," *Carbon,* vol. 49, no. 6, pp. 1894 - 1899, 2011.

[10] H. Zhu, Y. Zhan, L. Yue, W. Li, G. Li, D. Shu and H. Chen, "Graphite-Carbon Nanotube Composite Electrodes for All Vanadium Redox Flow Battery," *Power Sources,* no. 184, 2008.

[11] S. and I. , "Optimasi Deasetilasi Kitin Pada Udang," *Jurnal Kimia,* vol. 2, no. 5, pp. 9704-9730,

2009.
[12] S. Hirano, Chitin and Chitosan, Germany: Republicka of Germany, 1984.

IOP Conf. Series: Journal of Physics: Conf. Series **970** (2018) 012007 doi:10.1088/1742-6596/970/1/012007

The Influence of Red Fruit Oil on Creatin Kinase Level at Maximum Physical Activity

Fajar Apollo Sinaga[1*)]**Pangondian Hotliber Purba**[1]
[1]Faculty of Sports Sciences, State University of Medan, Indonesia.

*)Corresponding author: Email: sinaga_fajar@yahoo.com

Abstract. Heavy physical activities can cause the oxidative stress which resulting in muscle damage with an indicator of elevated levels of Creatin Kinase (CK) enzyme. The oxidative stress can be prevented or reduced by antioxidant supplementation. One of natural resources which contain antioxidant is Red Fruit *(Pandanus conoideus)* Oil (RFO). This study aims to see the effect of Red Fruit Oil on Creatin Kinase (CK) level at maximum physical activity. This study is an experimental research by using the design of randomized control group pretest-posttest. This study was using 24 male mice divided into four groups, the control group was given aquadest, the treatment groups P1, P2, and P3 were given the RFO orally of 0.15 ml/kgBW, 0.3 ml/kgBW, and 0.6 ml/kgBW, respectively, for a month. The level of CK was checked for all groups at the beginning of study and after the maximum physical activity. The obtained data were then tested statistically by using t-test and ANOVA. The result shows the RFO supplementation during exercise decreased the CK level in P1, P2, and P3 groups with $p<0.05$, and the higher RFO dosage resulted in decreased CK level at $p<0.05$. The conclusion of this study is the Red Fruit Oil could decrease the level of CK at maximum physical activity.

1. Introduction

Severe physical activity may result in oxidative stress resulting from the production of free radicals exceeding the amount of antioxidants in the body [1-2]. Under conditions of oxidative stress, free radicals will cause cell membrane lipid peroxidation [3]. Malondialdehyde (MDA) is one result of lipid peroxidation, so MDA is a common indicator used to determine the amount of free radicals and indirectly assess the body's oxidant capacity [4-5].

The results showed that maximal physical activity could lead to lower antioxidant levels of superoxide dismutase (SOD), Catalase (CAT), Glutathid Peroxidase (GPx) and glutathione-S-transferase in gastrocnemius [6]. The results of research conducted by Moflehi et al (2013) reported with increased intensity of exercise can lead to increased levels of MDA and creatin kinese (CK). It is known, CK is one indicator of the occurrence of damage from muscle cells (Koch, 2014). CK enzyme levels increase after high intensity exercise caused by damage to sarcolemma due to continuous movement in high intensity. Sarkolema damage causes the release of CK enzymes from muscle cells to the circulatory system so that this can be used as an indicator of muscle damage (Tortora 2009). Increased index of muscle damage associated with decreased performance (White et al. 2008). Muscle damage occurs when the muscles receive excessive physical stimuli. In exercise and exercise, muscle damage is caused by physical trauma, resulting in tearing, bruising, or rupture of muscle fibers, and miofilament disorders (Nosaka 2007).

IOP Conf. Series: Journal of Physics: Conf. Series **970** (2018) 012007 doi:10.1088/1742-6596/970/1/012007

In the body there is a mechanism of antioxidants or anti-radical free endogenically. The free radicals formed will be neutralized by the elaboration of the defense system between antioxidant enzymes and a number of non-enzyme antioxidants [2,8]. Research results reported that administration of antioxidants derived from natural or synthetic sources from outside the body is needed to neutralize the free radicals that are formed during physical activity especially heavy physical activity [9-12]. The results of a study conducted by Poulab et al (2015) showed that vitamin C supplementation for four weeks significantly decreased serum CK after eccentric exercise. The results of research conducted by Zoppi et al (2009) reported that supplementation of vitamin C and E in soccer players during undergoing exercise program can reduce the occurrence of lipid peroxidation and muscle damage but can not improve performance.

One natural source that is known to contain antioxidants such as beta carotene, tocopherol and fatty acids is a red fruit (Pandanus conoideus Lam) found in Papua [13-14]. As an antioxidant RFO is believed its ability to prevent chronic diseases such as cardiovascular, atherosclerosis, and cancer. In addition, Papuans believe that red fruit can increase stamina, but it still needs to be scientifically proven.

The results of research conducted by Sinaga (2015) reported that red fruit oil can increase the amount of erythrocytes, hematocrit and hemoglobin and decrease the number of rat leukocytes during the maximum physical activity. In his study also reported that the use of red fruit oil can decrease MDA and increase the resistance of mice when performing maximum physical activity [15]. The results of this study is very interesting so it needs to be examined further about the effects of red fruit oil in humans especially on CK levels in maximum physical activity.

The targeted finding/innovation is the discovery of one of the antioxidants to prevent or reduce free radicals due to physical activity so as to improve athlete performance and athletic health.

The results of the study are expected to contribute to the development of science and technology, especially as a basis for further research and development of phytopharmaca for improving public health, especially for athlete's health

The results of this study can be applied to athletes during an exercise program or at the time of the competition so as to support development programs in the field of sports achievements and health. In terms of science and technology development, this research is one form of disciplinary contribution other than the discipline of sports to support the performance of athletes.

2. Methods

The research will be conducted at Animal Laboratory Unit of Biology Division, Faculty of Natural Sciences State University of Medan for 6 weeks. The examination of CK levels was conducted at the Integrated Laboratory of the Faculty of Medicine, University of North Sumatra.

Implementation of Research

The number of male rats used in the study were 24 tails. Selected healthy male rats and performed adaptation for 7 days. Then the rats were divided into 4 groups randomly. Each group consists of 6 rats. P0 (control group) was given 2ml water, group P2, P3 and P4 were given MBM respectively 0.15ml, 0.3ml and 0.6ml / day using Gavage syringes. Next all the rats were made swimming exercises for a month. In the first week for 15 minutes / day, Sunday II 20 minutes, Sunday III 25 minutes and week IV 30 minutes. After 30 days, all rats were ordered to perform maximum physical activity by means of rats forced swimming in a container with no way out. Further measurements of CK levels and duration of swimming mice (endurance).

3. Results

Data of CK levels before treatment and after treatment in each group are shown in Table 1.

Table 1. Average CK Rate Before and After MBM In Maximum Physical Activity

Level	Treatment	P0	P1	P2	P3	F	p
CK	Before	128, 93	128,67	128,33	128,40	0,505	0,683
	After	226,82	203,51	166,89	129,71	2792,344	0,000
t dependent		0,000	0,000	0,000	0,000		

Data of CK levels both before treatment and after treatment in each group were tested for normality were tested for normality by using Shapiro-Wilk test and homogeneity with Levene's test. The result shows that all data are normal and homogeneous distribution (p> 0,05), so parametric test, One Way Anova test and t-paired test.

Analysis of significance with One Way Anova test of CK content before administration of MBM showed that F = 0,505 and p = 0,683. This means that the four groups of CK average are not significantly different (p> 0.05). Analysis of significance with One Way Anova test of CK level after MBM administration during one month of training program showed that F = 2792,344 and p = 0,000. This means that the four groups of mean CK differed significantly (p <0.05).

Based on t-paired test, it was found that there was an increase of CK levels in the control group (P0), whereas in group P1, P2 and P3 decreased significantly (P <0.05).

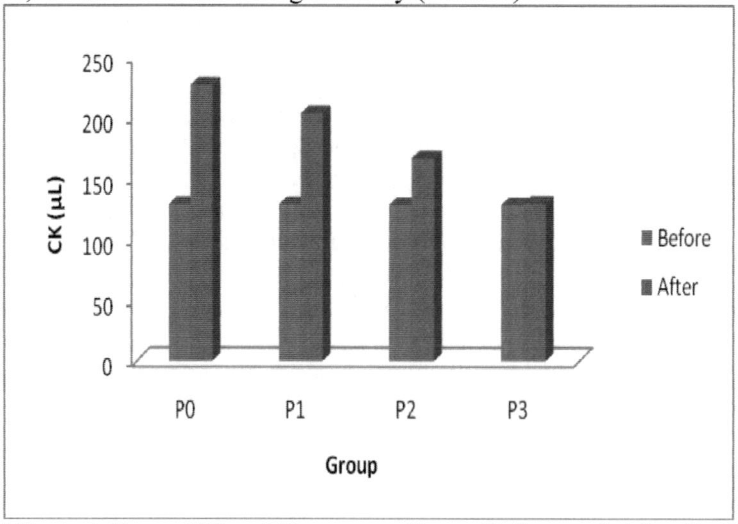

Figure 1. Level of Control Group CK and Treatment Group Before and After Performing Maximum Physical Activity

4. Discussion

Decrease Creatin Kinase (CK) Content After Red Fruit Oil Based on the analysis result, it was found that the mean of Creatin Kinase (CK) of control group was 128.93, mean of MBM group dose 0,15 ml/day was 128.67 dose 0,3 ml/day was 128.33 and the dose of 0.6 ml/day was 128.40. Analysis of significance with One Way Anova test showed that the mean Creatin Kinase (CK) content in all four groups before treatment was not different (p> 0,05). Furthermore, after treatment, based on the analysis result showed that the average Creatin Kinase (CK) of control group was 226.82, the mean of MBM group dose 0,15 ml/day was 203.51, dose 0,3 ml/day is 166.89, and the dose of 0.6ml/day is 129.71. Based on the results of the analysis it was found that after treatment between the four groups the results differed significantly (p <0.05).

In normal circumstances, free radicals form slowly, then neutralized by the antioxidants present in the body. But if the rate of free radical formation is greatly increased because triggered by excessive exercise, the amount of free radical will exceed the ability of the body's defense system and can not be neutralized by antioxidants in the body. This can result in damage to the cell membranes, muscles, bones and tissues [16].

The increase in Creatin Kinase (CK) levels after being given maximum physical training, can cause physical stress. For that it takes a recovery period that is the time it takes the body to return to its original state from the state of the training activity [17]. Increased oxygen consumption by the body during heavy exercise may increase ten to twenty times or more. Under high stress, in muscle fibers there is an increased use of oxygen above the normal requirement. This extraordinary increase in oxygen can trigger the release of free radicals in the mitochondria, resulting in increase CK use in the body. This causes the number of CK in the body to increase. While in the non-contracting muscles, the amount of oxygen will decrease which causes the ischemic process. With antioxidant supplementation it is expected that CK count will decrease [16].

In this study, administration of red fruit oil can decrease CK levels. The decrease in CK levels is due to red fruit oils rich in natural antioxidants such as 11,500 ppm carotenoids, beta carotene 694,80 ppm, total tocopherol 11,200 ppm and α-tocopherol 495.50 ppm [14].

As one source of natural antioxidants, antioxidant activity of Red Fruit Oil has been investigated. In a study conducted by Rohman et al (2010) reported that red fruit oil had antioxidant activity on in vitro assay with IC50 value of 451.51 μg / ml. In in vivo testing, red fruit oil dose 0.15 ml / kgBW; 0.3 ml / kg BW and 0.6 ml / kg BW may decrease MDA levels in the blood [15,19]. Malondialdehyde is one result of lipid peroxidation [4,5,15], so MDA is a common indicator used to determine the amount of free radicals and indirectly assess the oxidant capacity of the body.

The results of this study were supported by the results of a study conducted by Sandhiutami (2012) who examined the determination of tocopherol content in red fruit oil in blood in male wistar rats subjected to maximum physical activity load. In the study, measurements of tocopherol content in the blood of rats fed red fruit oil with different doses (0.15ml / kgBW, 0.3ml / kg / BW and 0.6ml kg / BW). The results of measurements show that red fruit oil in rats may increase levels of tocopherol in rat blood after maximum physical activity [19].

The results of a study conducted by Armiyanti et al (2009) who examined the effect of red fruit oil with various concentrations on endothelial cells exposed to serum of severe malaria patients and normal neutrophils concluded red fruit oil can decrease oxidative stress characterized by decreased production of Reactive Oxygen Intermediates (ROI) in the treatment group with red fruit oil concentration 2.8%, 5.7%, and 11.3%. In the treatment group treated with red fruit oil the concentrations showed higher concentrations of less endothelial ROI cell production, even at the highest concentration (11.3%) of ROI production near zero [20].

5. Conclusion

Red Fruit Oil supplementation during exercise program can increase the level of Creatin Kinase at maximal physical activity

Acknowledgements

The authors thank to Directorate for Higher Education, The National Education Ministry of Republic Indonesia for for Research Grant of Hibah Fundamental for financial support.

Reference

[1] Daniel, R.M., Stelian, S., Dragomir, C. (2010), The effect of acute physical exercise on the antioxidant status of the skeletal and cardiac muscle in the Wistar rat. *Romanian Biotechnological Letters*. Vol. 15, No. 3, Supplement, p 56-61.

[2] Urso, M.L., Clarkson, P.M. (2003), Oxidative stress, exercise, and antioxidant supplementation. *Toxicology* 189(1-2):41-54A

[3] Evans, W. J. (2000), Vitamin E, vitamin C, and exercise. *Am J Clin Nutr,* 72, 647S-52S.

[4] Wang, L., Zhang, H.L., Zhou, Y.J., Ma, R., Lv, J.Q., Li, X.L., Chen, L.J. and Yao, Z. (2008). The decapeptide CMS001 enhances swimming endurance in mice. *Peptides*, 29: 1176-1182.

[5] Lyle, N., Gomes, A., Sur, T., Munshi, S., Paul, S., Chatterjee S. and Bhattacharyya, D. (2009). The role of antioxidant properties of Nardostachys jatamansi in alleviation of the symptoms of the chronic fatigue syndrome. *Behavioural Brain Res.*, 202: 285-290.

[6] Thirumalai, T., Viviyan T. S., Elumalai,E.K., David, E. (2011). Intense and exhaustive exercise induce oxidative stress in skeletal muscle, *Asian Pacific Journal of Tropical Disease* 63-66

[7] Koch, A.J., Pereira,R., Machado,M. (2014). The creatine kinase response to resistance exercise. *J Musculoskelet Neuronal Interact.* 14(1):68-77

[8] Christopher, P.I.,Wenke, J.C., Nofal,T., Armstrong, R.B. (2004), Adaptation to lenghthening contraction-induced injury in mouse muscle. *J.Appl.Physiol* 97:1067-76.

[9] Ostojic, S.M., Stojanovic, M.D., Djordjevic, B., Jourkesh, M., Vasiljevic, N. (2008). The Effect of a 4-Week Coffeberry Suplementation on Antioxidant Status, Endurance, and Anaerobic Performance in College Athletes. *Research in Sports Medicine,* 16: 281-294

[10] Gupta, C., Gupta, P.H and Singh, B. (2009) Effect of Vitamin Supplementation on Exercise Induced Oxidative Stress in Trained Elite Indian Cyclists. *Am. J. Biomed. Sci.* 1(2), 166-170

[11] Bentley, D.J., Dank,S., Coupland,R., Midgley,A., Spence,I., (2012), Acute Antioksidant Suplementation Improves Endurance Performance in Trained Athletes, *Research in Sports Medicine,* 20: 1-12.

[12] Kalpana, K., Kusuma, D.L., Lal, P.R., Khanna, G.L. (2012). Effect of Spirulina on Antioxidant Status and Exercise-Induced Oxidative Stress of Indian Athletes in Comparison to a Commersial Antioxidant. *Asia Journal of Exercise & Sports Science.* Vol 9 (No.2); 36-48

[13] Budi I.M., Buah Merah. Jakarta: Penebar Swadaya; 2005.

[14] Alamsyah, A. N. (2005). *Perpaduan sang penakluk penyakit VCO + buah merah.* Agromedia Pustaka: Jakarta, hal: 2-45

[15] Sinaga,F.A. (2015). The Effect of red fruit oil on rat's malondialdehide at maximal physical avtivity. *Proceeding international conference of ASEAN Council of Physical Education and Sports (ACPES) 2015.* ISBN 978-979-1964-4-2. P 379-385

[16] Cooper, K. 2001. Sehat Tanpa Obat. Empat Langkah Revolusi Antioksidan yang Mengubah Hidup Anda. Cetakan Pertama. Bandung: Kaifa.

[17] Judge, S., Mok Jang, Y., Smith, A., Hagen, T., dan Leeuwenburg, C. 2005. Age-Associated Increases in Oxidative Stress and Antioxidant Enzyme Activities in Cardiac Interfibrillar Mitochondria: Implication For the Mitochondrial Theory of Aging. University of Florida. *The FASEB Journal.*

[18] Rohman, A., Riyanto, S., Yuniarti, N., Saputra, W.R., Utami, R. and Mulatsih, W. 2010a. Antioxidant activity, total phenolic, and total flavonoid of extracts and fractions of red fruit (*Pandanus conoideus* Lam). *International Food Research Journal* 17: 97-106.

[19] Sandhiutami,NMD., Ngatijan., Kristin,E. (2012). Quantitative assay for tocoferol of Buah Merah (*Pandanus conoideus* LAM) oil in rat's blood given maximum phisical activities

[20] Armiyanti,Y., Fitri,L.E., Widjajanto,E. , (2007). The Effect Of Red Fruit Oil (*Pandanus Conoideus)* On Oxidative Stress Of Endothelial Cells Exposed To Malaria Falciparum Patient Serum And Healthy Donor Neutrophil**.** Jurnal Kedokteran Brawijaya, Vol. XXIII, No. 1, April 2007

IOP Conf. Series: Journal of Physics: Conf. Series **970** (2018) 012008 doi:10.1088/1742-6596/970/1/012008

The Activity Test of Ethanol Extract Tea Parasite Herb (*Scurrulla Artopurpurea*) as an Immunostimulator on Wistar Strain Rat Sensititized with Sheep Red Blood Cell

Retno Yulianti[1a], M.Dimas A[2], Aldi Hafiz[3] and Muttia Amalia[4]

[1]Departement of Anatomy-Pathology FK UPN "Veteran"Jakarta [4]Departement of Clinical Pathology FK UPN "Veteran" Jakarta [2,3]UndergraduateStudent ,Faculty of Medicine,UPN "Veteran" Jakarta Jl.RS Fatmawati, PondokLabu Jakarta Selatan 12450

[a]Corresponding author's: dr.retnoyulianti@yahoo.com

Abstract. Tea parasite (Scurrulla artropurpurea) is one of the hemiparasite plants suspected as an imustimulator agent. The aims of this study was to prove the activity of tea parasite herb extract as immunostimulator in Wistar strain rats that are sensitized by sheep red blood cell suspension (SRBC). Twenty eight Wistar rats were divided into 4 groups. The control group was given CMC 1%, the treatment group was given the tea parasite herb extract at doses of 750 mg, 1.5g and 3 g/kgBW/day, each group was sensitized by SRBC 10% as much 0.5 ml intraperitoneally on the 0th and 9th days and 0.1 ml intraplantar on 12th days. The total number of leukocyte and lymphocyte cells was performed using blood smear and analyzed by GLMRM. test. The macrophage cell activity and the capacity of the peritoneal fluid preparation on day 14 and analyzed by ANOVA test. The histopathological features are demonstrated after 48 hours of intraplantar injection, quantitatively perceived perivascular and periadnexal infiltrates. The results showed that there were average differences on leukocyte cell counts (F = 46.249) and total lymphocyte cells (F = 58.144) on each calculation (day), but there were no average differences between each treatment group. The highest activity of phagocytosis is 78% and continues to increase along with the additional dose of extract, without the increased capacity of macrophage phagocytosis. Histopathological features of slow-type IV reaction shows mild severity. It can be concluded that the active compound of the extract tea parasite herbare able to improve immune system.

1. Introduction

Indonesia is a part of a tropical country. Areas that have a tropical climate are perfect for the growth of various disease agents such as bacteria, viruses and so on. This is what facilitates the immune system weakened, resulting in various diseases [1]. Humans since birth have been equipped with a specific and non-specific body defense system. The immune system works to counteract various bacteria, viruses, fungi, and other foreign substances that can cause various diseases [2]. If interference occurs by the antigen (foreign body), then the body's defense system will be activated to maintain the body against infection of microorganisms, homeostasis against the elimination of old components of the body and supervision against the destruction of cells that mutate, especially the malignant [3].

Immunostimulator is a substance capable of improving the function and activity of the immune system, so as to restore the imbalance of the impaired immune system by stimulating and improving the function of it. Both immune systems work simultaneously in eliminating the antigens that enter the body. Materials that can stimulate immune cells can be from biological or synthetic groups [4]. Immunostimulators relate to the presence of antigens that enter our body and are able to strengthen the immune system naturally by inducing immune responses through complementary phagocytosis, IgA antibody secretion, release of α and β interferons, increasing T and B lymphocytes, synthesizing specific antibodies [5]

Tea parasite (Scurrulla artropurpurea) is one of the hemiparasite plants suspected as an imustimulator agent that has 16 bioactive compounds such as polysaccharide compounds, steroids / triterpenoids, alkaloids, saponins, tannins, quinones and flavonoids [6,7]. Benalu has long been used as a cure for various diseases, according to Winarno et al. (2000) that tea parasites have. Infusion of tea parasite is suspected to have immunostimulator activity in cancer cells in C3H mice by increasing the concentration of immunoglobulin G (IgG). The immunostimulator effect is due to the content of quercetin contained in the parasitic extract of tea. Quercetin is thought to induce the proliferation and maturation of macrophages and is able to induce T lymphocytes or lymphoid maturation [7]. In addition, the Lectin content contained in tea parasite is cytotoxic and able to boost the immune system [8]. Polysaccharides contained in the parasite can increase the secretion of antibodies and cytokines, either by increasing the function of Natural Killer cells and T and B lymphocytes [9].

In this research, immunostimulator activity of ethanol extract of parasite (Scurrulla artropurpurea) in white male Wistar strain is first induced by sheep red blood cells as antigen.

2. Research Methods

The design of this study used a preposttest experimental method with a control group, conducted in the laboratory using tea parasite test (Scurrulla atropurpurea) materials and white male rat (Rattus norvegicus) strain of male Wistar strain previously induced with sheep red blood cells, then blood sampling for the total number of lymphocytes, leucocyte cells, activity and capacity of phagocytosis and histopathologic features as a quantitative type of slow hypersensitivity immune response.

The ethanol extract of tea parasite was obtained from BALITRO, Bogor using the maceration method and the phytochemical simplicia has been tested that includes examination of secondary metabolites consisting of alkaloid compounds, flavonoids, saponins, tannins, steroids / triterpenoids.

The ethanol extract of tea parasite was suspended with 1% Na CMC 2 ml, then administered to the orally daily rats for 14 days with dose of 750 mg / kgBW / day, 1.5g / kgBW / day and 3g / kgBW / day.

Animal for testing is done by ethical approval by Medical Ethics Committee of Faculty of Medicine UPN Veteran Jakarta with approval number B / 1156 / VII / 2017 / KEPK. Prior to the experiment, Wistar strain rat were adapted for one week and each group of rats was kept in a cage, fed standard feeding ad libitum. Then each group was sensitized with 10% of red blood cell sheep (SRBC) suspension intraperitoneally on days 0 and 9 and 0.1 ml intraplantar on day 12 [10]. The samples were taken from animal experiments with a minimum sample size of 28 tails calculated on the Faraday formula. Then divided into 4 groups namely 1 control group given 2 ml CMC 1% and 3 treatment groups.

According to Nurmaya Effendi (2014), 1 mL of fresh sheep blood is accommodated in a tube containing 1 mg EDTA as an anticoagulant. Then centrifuged 3000 rpm for 10 minutes. After forming two layers of plasma and red cell deposits, then the plasma fluid is removed by using a micropipette. Then the precipitate was washed with PBS pH 7.2 in the tube three times. This procedure is repeated until the top layer is completely clear and colorless. The clear top layer is removed and the bottom layer is a 100% SRBC suspension. Take 0.5 mL of 100% SRBC suspension, add PBS with same volume, so get 50% SRBC suspension. Prepare the antigen to be used by diluting 0.4 mL of 50% SRBC suspension with 1.6 mL PBS to obtain 2 mL of antigen suspension (SRBC 10%) [11].

The white rat was taken by the ventral part of the vein by the tail section compressing warm water for 5 minutes in advance to dilate the vein, then the rat tail was cut. Blood that came out immediately taken and accommodated on a 1.8 ml effendrof tube which drops with EDTA 4% 2 drops. The pipette is then shaken for three minutes until homogeneous. Blood sampling was performed on days 0, 7 and 14.

Take the blood from the tube by inhalation using a pipette until the mark 0.5 and the tip of the pipette is cleaned with a tissue. Then suction the Turk solution with the same pipette until it reaches the limit of the number 11. Before the solution is put into the count chamber, remove the first two or three drops. After the solution is put into the counting chamber awaited for one minute, the next leukosit is calculated using magnification 10 times or 40 times on the objective lens. The leukocyte count formula is number of cells x 50.

Total lymphocyte counts were performed by using the multiplication of leukocyte count and lymphocyte count [3, 12]. The calculation of the absolute number of each leukocyte type is calculated by the formula is the number of cell of the leukocytes type :100) x total number of leukocytes

On the 14th day, the euthanized rat was then dissected in the stomach by using surgical scissors and sterile tweezers. Intraperitoneal fluid is taken using a micropipette. Then the intraperitoneal fluid is applied to the object glass and fixed with methanol for 5 minutes with Giemsa staining. Then let stand for 20 minutes and then flowed with running water. After dry preparation of the dosage under a microscope and with the added emergent oil with 10x-100x enlargement is calculated the activity and capacity of macrophage phagocytosis. Phagocyte activity is determined by percentage of the number of phagocyte cells that will perform phagocytosis process in 100 phagocyte cells. The phagocyte capacity is based on the amount of SRBC to be in phagocytosis by 50 active phagocyte cells [13].

Sensitized rats with intraperitoneal SRBC suspension were injected with an intradermal 0.1 ml SRBC10% on the sole of their right foot on day 12. Injecting effects were seen after 48 hours of exposure to sheep red blood cells, then rats in anesthesia with later ether rats were killed by means of cervical dislocations, then taken the soles of their feet to be processed into histologic preparations after paraffin blocks were made. Values for inflammatory severity are seen separately for periadnexal and perivascular inflammation for each section. Each part of the skin was given an overall score based on the number of points for periadnexal and perivascular inflammation [14]

3. Research and Discussion Result

The result of phytochemical simplicia test from qualitatively tea extract by BALITRO laboratory, Bogor was found to contain positive alkaloid, saponin, tannin, phenolic, flavonoid, triterpenoid, steroid and glycoside compounds. This is consistent with a study by Simanjuntak (2004), that tea parasites in the genus Scurrula contain 16 important compounds, six unsaturated fatty acid compounds, two xantins, two flavonol glycosides, one monoterpene glycoside, one lignin glycoside and four flavones [15] .

The test animals used were male rats aged 3-4 months, according to the adult age in humans, so that the expected hormone levels in the body have been stabilized and the immune system in male rats is less affected by the reproductive hormone. Basically the human's and rats' immune system are almost the same [16].

Table 1. Result of Calculating Average Number of Leukocyte Cells and Total Lymphocyte Cells in Wistar Wistar Tissue Between Treatment Group Compared to Control Group on Day 0, 7, and 14 (Thousand / mm3)

	Control Group		Treatment Group 1		Treatment Group 2		Treatment Group 3	
	Leukosit	Limfosit total	Leukosit	Limfosit total	Leukosit	Limfosit total	Leukosit	Limfosit total
day-0	12.76	7.79	15.36	10.51	13.86	9.38	12.19	7.75
day-7	23.22	16.68	24.11	18.27	23.07	17.88	18.46	13.87
day-14	6.69	4.22	8.69	5.18	7.58	5.60	7.72	4.67

The first intraperitoneal sheep red cell sensitization will stimulate non-specific immune systems, i.e leukocyte cells to eliminate pathogens, although the body has not been exposed before, whereas the specific immune system will be triggered after the body gets the same previous antigen exposure through memory immune cells. Specific immune cells play primarily lymphocyte cells, because they have the main characteristics of specificity, disversity, memory, specialization and self-limiting [17]

The result showed that the average number of leukocyte cell and total lymphocyte cell of Wistar strain on each treatment on days 0, 7, and 14 as shown in Table 1.

The average difference of the total lymphocyte cells number in Wistar rats on day 0, 7th and 14th day can be known by the *Within subjects contrasts test* as shown in Table 3 and differences between treatment groups can be obtained by testing the *Between subjects effects* as listed in Table 4 based on analysis general linear test model repeated measures (GLM-RM) test.

Table 2. Within-Subjects Contrasts Test Results

Source	Day	df	F	Sig.
Day	Level 1 vs. Level 2	1	58.144	.000
	Level 2 vs. Level 3	1	225.137	.000
Day * treatment	Level 1 vs. Level 2	3	0.355	.786
	Level 2 vs. Level 3	3	1.236	.319
Error(Day)	Level 1 vs. Level 2	24		
	Level 2 vs. Level 3	24		

Table 3. Between subject effects Test Results

Source	df	F	Sig.
Intercept	1	527.078	0.000
Perlakuan	3	1.825	0.17
Error	24		

Information:
- Level 1 = day 0, Level 2 = Day 7, Level 3 = Day 14
- Treatment = control group (placebo CMC 1%), treatment (tea parasite extract 750 mg, 1.5g, 3g / kgBB / day)

.

The average difference of the leukocyte cells number in Wistar rats on day 0, 7th and 14th day can be known by the *Withn subjects contrasts test* as shown in Table 4 and differences between treatment groups can be obtained by testing the *Between subjects effects* as listed in Table 5 based on analysis general linear test model repeated measures (GLM-RM) test.

Table 4. Within-Subjects Contrasts Test Results

Source	Day	df	F	Sig.
Day	Level 1 vs. Level 2	1	46.249	.000
	Level 2 vs. Level 3	1	313.686	.000
Day * Treatment	Level 1 vs. Level 2	3	0.474	.703
	Level 2 vs. Level 3	3	2.485	.085
Error(Hari)	Level 1 vs. Level 2	24		
	Level 2 vs. Level 3	24		

Table 5. Between subjects effectsTest Results

Source	df	F	Sig.
Intercept	1	611.742	.000
Treatment	3	1.348	.282
Error	24		

Information:
- Level 1 = day 0, Level 2 = Day 7, Level 3 = Day 14
- Treatment = control group (placebo CMC 1%), treatment (tea parasite extract 750 mg, 1.5g, 3g / kgBB / day)

On day 0, the amount of leukocyte cell is within the normal range that is 4,5x103-11x103 cell/μL. On day 7, after a desensitizen of SDMD antigen, a respond of non specific immune increases the amount of leukocyte cells. This indicate that antigen difused to the vascular, continues to be phagocytosised by leukocyte cells. Limfocyte cells appear on day 5 and reach the maximum amount on day 7. According to theory, the increasing amount of limfocyte cells are caused because of

IOP Conf. Series: Journal of Physics: Conf. Series **970** (2018) 012008 doi:10.1088/1742-6596/970/1/012008

proliferation dan the differences are triggered by new antigens that is shown by APC. Later on day 14, there is a tendency of the decreasing amount of leukocyte cells and limfocyte cells because the extract of Benalu tea intensify the immune response on a limited time, with a limited effect. This happened because it depends on the amount of dose tested, if it is high it will cause imunosupression and if it is low it will cause imunostimulation. Furthermore, the technique of dose tested will affect the strength of immune response that appears. This immunostimulation ability seems to be related to polysaccharide content in parasite tea extract which can increase the secretion of antibodies and cytokines by stimulating the function of natural killer cells and also T and B lymphocytes.

Macrophages have a very important role as phagocytes in the innate immune system and as antigen presenting cells (APC) that initiate an adaptive immune response. TLR can activate innate immune responses and stimulate the production of various proteins that play a role in the important function of macrophages. From the results of the research, the calculation of activity and capacity macrophage phosphocytic of Wistar strain rats on each treatment as shown in Table 6 and 7.

Table 6. Results of Fagocytosis Macrophage Rats Wistar Triangle Tests

Group	Macrophage Activity
Control	70%
Treatment 1	71%
Treatment 2	69%
Treatment 3	78%

Table 7. Results Calculation of Meaning of Intergroup Macrofag Fagocytosis Capacity

Group	Kapasitas Fagositosis
Control	567
Treatment 1	310
Treatment 2	216
Treatment 3	371

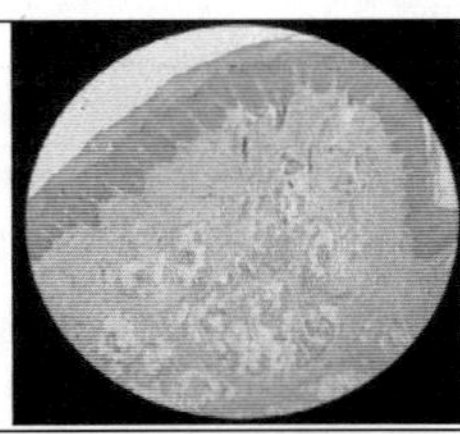

Group 1 : inflammatory cells are located in small numbers around glandular and follicular structurs, some inflammatory cells may be present in the superficial dermis. No infiltrate perivascular inflammatory

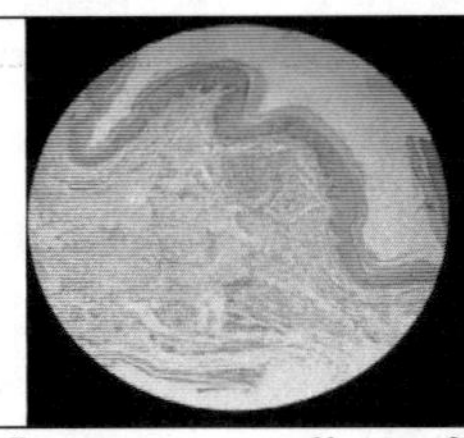

Group 2 : inflammatory cells cuffs some adnexal structures, small numbers inflammatory cells scattered in the superficial dermis. Inflammatory cells are located around vessels and scatteres around these areas, they may be present in small numbers around adipocytes in the hypodermis, present around the majority of vessels

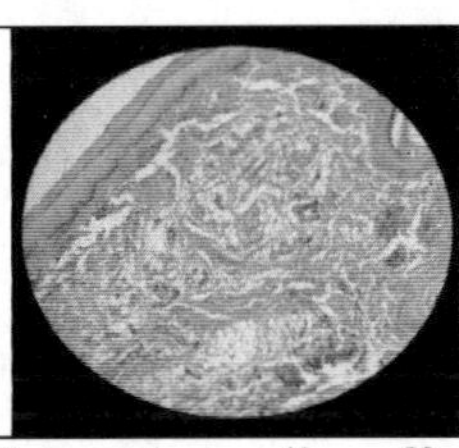

Group 3 : inflammatory cells cuffs some adnexal structures, small numbers inflammatory cells scattered in the superficial dermis. Inflammatory cells are located around vessels and scatteres around these areas, they may be present in small numbers around adipocytes in the hypodermis, present around the majority of vessels

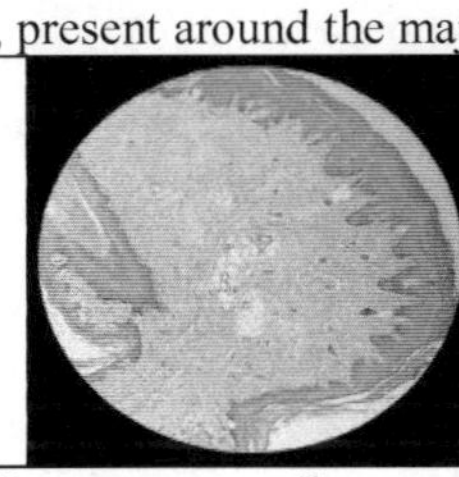

Group 4 : inflammatory cells are located in small numbers around glandular and follicular structurs, some inflammatory cells may be present in the superficial dermis. No infiltrate perivascular inflammatory

Figure 1. Histopathology of the skin as a response to slow type hypersensitivity

Statistical test results using One-Way ANOVA and showed significance value ($p < 0.05$) showed that there was a significant difference in macrophage phagocytosis activity between groups and no significant ($p > 0.05$) difference in macrophage phagocytosis activity between groups. Further statistical test results with Post Hoc test to see comparison between groups showed similar results with the Anova test.

According to the results of macrophage measurements, Benalu tea extract is not high enough to increase the capacity of macrophage phagocytosis, even though with the maximum additional dose. The differences taken of the activity of macrophage fagocythosis in this research is allegedly caused by presence of flavonoid and saponin contained in the Benalu tea extract. Francis *et al* [18] stated that saponin increases the macrophage fagocytosis. Flavonoid potentially works on limfokin that is produced by T cell, so it will stimulate the phagocyte cells that responds to phagocytosis [19]. As an imunostimulan, flavonoid and saponin compound are used to improve the immune system by increasing the phagocytosis by granulocyte and macrophageand stimulate the production of sitokin by Th2 cells so the production of antibodies will increase [20].

Delayed Type hypersensitivity (DTH) response is a reaction of increased reactivity and sensitivity to antigens previously exposed and used to determine the activation of T cells involved in cell-specific immune responses. Histopathologically can be seen the existence of inflammatory response, existing cell types, inflammatory severity and location of infiltration is by assessing periadneksal and perivaskular infiltrates (Figure 1).

Overall, histologic features do not show a significant effect of DTH response. This is possible because flavonoid compounds can affect complex immune responses in complex and biphasic action, ie low doses can stimulate lymphocyte proliferation but otherwise high doses preclude it.

4. Conclusions

Overall, test animals treated with tea extracts at various doses did not show any significant difference, so further studies of other immunity parameters were necessary.

Acknowledgements

The authors would like to thank LPPM UPN Veteran Jakarta and Kemenristek Dikti for the Beginner Lecture Grant which has provided financial support to this research.

References

[1] Sukowati S. (2010). Masalah Vektor Demam Berdarah Dengue (DBD) dan Pengendaliannya di Indonesia. Buletin Jendela Epidemiolog Vol.2

[2] Tjokronegoro A. (19990). Peranan Mikrooganisme dan Komponennya Sebagai Imunomodulator Sistem Imun Tubuh Manusia. *Majalah Kedokteran Indonesia* 1990; 40(11). hal.613-19.

[3] Baratawidjaja, K G. 2009. *Imunologi Dasar*. Jakarta: Fakultas Kedokteran Universitas Indonesia

[4] Wiedosari, E. 2007.Peranan Imunologi Alami (*Aloe vera)* dalam Sistem Imunitas Seluler Humoral.. *Watazoa*; Vol. 17 (4)

[5] Petrunov B., Nenkov P., and Shekerdjiisky R. (2007). *The role Of Immunostimulants In Immunotherapy And Immunoprophylaxis*. Bulgaria: National Center of Infectious and Parasitic Disease, BulBio-NCIPD, Natsim Ltd

[6] Wagner H., dan Jurcic K. (2009). *Assay for Immunomodulation and Effect on Mediators of Inflamation*, Dalam: Dey, P. M., dan Harbone, J. B., (Eds.), *Methods in Plants Biochemistry: Assay for Bioactivity.* Edisi Keenam, Academic Press, London cit.

[7] Winarno M W., D. Sundari dan B. Nuratmi. (2000). Penelitian Aktivitas Biologik Infus Benalu Teh (Scurrula atropurpurea Bl. Dans) Terhadap Aktivitas Sistem Imun Mencit. *Cermin Dunia Kedokteran*.2000; Vol.127: 11-14.

[8] Barnes J, L A Anderson and J D Philipson. (2002). *Herbal Medicine*. 2nd ed.. Pharmaceutical Press. Great Britain

[9] Chen HL, Li DF, Chang By. (2003). Effec of Chinese Herbal Polysaccharides on the Immunity and Growth Perforance of Young Broilers. *Poult Sci* 82:364-370

[10] Anggraini, D R. (2012). Ekspresi Imunohistokimia Ki-67 pada Tumor Payudara Tikus Wistar yang Diinokulasi Kanker Terinduksi Benzo (α) Pyrene dengan Pemberian Ekstrak Benalu Teh. *Tesis*. Universitas Sumatra Utara

[11] Effendi N dan Widiastuti H. (2014). Identifikasi Aktivitas Imunoglobulin M (Ig.M) Ekstrak Etanolik Daun Ceplukan (Physalis Minima Linn.) Pada Mencit. *Jurnal Kesehatan*. 2014; Vol.7(2)

[12] Patel S, Banji D, Banji J F O, Patel M M, Shah K K. (2010). Scrutinizing the Role of Aqueous Extract of Trapa Bispinosa as an Immunomodulateor in Experimental Animals. *Int J Res Pharm Sci*.1(1) : 13-19

[13] Herawati I; Husin, A Asep; Sudigdoadi S. (2015). Pengaruh Ekstrak Etanol Propolis Terhadap Aktivitas dan Kapasitas Fagositosis pada Kultur Makrofag yang diinfeksi *Enteropathogenic Escherichia coli* (EPEC). MKB; Vol..47(2): 102-108

[14] EL Gulliver, K. M. Plain, D. J. Begg and R. J. Whittington. (2015). Histopathological Characterization of Cutaneous Delayed-type Hypersensitivity and Correlations with Intestinal Pathology and Systemic Immune Responses in Sheep with Paratuberculosis. *J.Comp.Path*; Vol.153: 67-80

[15] Simanjuntak, P; Parwati, T; Lenny, L E; Tamat, S R dan Murwani, R (2004). Isolasi dan Identifikasi Senyawa Antioksidan dari Ekstrak Benalu Teh (*Scurrula oortiana* (Korth) Danser).Jurnal Ilmu Kefarmasian Indonesia.2004; Vol.2(1):19-24

[16] Kresno SB. 2010. *Imunologi : Diagnosis dan Prosedur Laboratorium*. Edisi kelima. Balai Penerbit Fakultas Kedokteran, Universitas Indonesia, Jakarta

[17] Sringingsih dan Wibowo, A. E.(2006), Efek Protektif Pemberian Ekstrak Etanol Herba Meniran (Phyllanthus niruri L.) Terhadap Aktivitas dan Kapasitas Fagositosis Makrofag Peritoneum Tikus, *J. Amcarpus*, 6 (2), 91 – 96.

[18] Francis G, Kerem Z, Makkar, HPS 2002, The Biological Action of Saponins in Animal Systems: a review. *British J Nutri*

[19] Indonesia. Depkes, 2000, Parameter Standar Umum Ekstrak Tumbuhan Obat, Jakarta

[20] Rajput ZI, Hu SH, Xiao CW, Arijo AG 2007, *Adjuvant effects of saponins on animal immune responses*. J Zhejiang

IOP Conf. Series: Journal of Physics: Conf. Series **970** (2018) 012009 doi:10.1088/1742-6596/970/1/012009

Development Instrument's Learning of Physics Through Scientific Inquiry Model Based Batak Culture to Improve Science Process Skill and Student's Curiosity

Derlina Nasution, Putri Syahreni Harahap, Marabangun Harahap

Physics Education Program, Graduate State University of Medan

putri.syahreni.hrp@gmail.com

Abstract. This research aims to: (1) developed a instrument's learning (lesson plan, worksheet, student's book, teacher's guide book, and instrument test) of physics learning through scientific inquiry learning model based Batak culture to achieve skills improvement process of science students and the students' curiosity; (2) describe the quality of the result of develop instrument's learning in high school using scientific inquiry learning model based Batak culture (lesson plan, worksheet, student's book, teacher's guide book, and instrument test) to achieve the science process skill improvement of students and the student curiosity. This research is research development. This research developed a instrument's learning of physics by using a development model that is adapted from the development model Thiagarajan, Semmel, and Semmel. The stages are traversed until retrieved a valid physics instrument's learning, practical, and effective includes :(1) definition phase, (2) the planning phase, and (3) stages of development. Test performed include expert test/validation testing experts, small groups, and test classes is limited. Test classes are limited to do in SMAN 1 Padang Bolak alternating on a class X MIA. This research resulted in: 1) the learning of physics static fluid material specially for high school grade 10th consisted of (lesson plan, worksheet, student's book, teacher's guide book, and instrument test) and quality worthy of use in the learning process; 2) each component of the instrument's learning meet the criteria have valid learning, practical, and effective way to reach the science process skill improvement and curiosity in students.

1. Introduction

The purpose of national education, among others, is a human who has the ability of science and technology with the attitude and behavior in accordance with the values of the Indonesian nation. In the list of laws no.20 of 2003 on the national education system which states that national education serves to develop the ability and build the character and civilization of the nation, the choice for the development of potential learners to be a human being who believes, piety to God Almighty, noble character , healthy, knowledgeable, capable, creative, independent and become a democratic and responsible citizen (Depag RI, 2006).

Meanwhile, Indonesia is a country with a diversity of ethnic cultures and religions, where indigenous Indonesians consist of more than 300 ethnic groups with unique cultural identity and language and when viewed from the geographical conditions of Indonesia as an archipelagic country, the population is scattered in the mountains, coastal areas or in mainland with diverse environmental characteristics

(Danoebroto, 2012). This shows that Indonesia is a country with a very multicultural population. However, the dynamics of plurality due to the current era of globalization have a negative impact on the culture of the Indonesian nation with the erosion of regional cultural values and the spirit of nationalism caused by the impact of cultural values adopted by Indonesian society with the values of culture from outside, the Indonesian nation to lose its identity (Danoebroto, 2012).

One of the efforts that can be done is through the school environment that takes into account the local culture to know each other and understand the cultural diversity so as to cause mutual respect for the same or different ethnic identity. The classroom is considered quite effective as a place to recognize and re-understand the traditions and cultures that exist in Indonesia. It is also in line with the purpose of national education, it is necessary components that must be considered by the teacher as a preparation of the need for the learning process in the classroom such as learning approaches, learning models, student characteristics, environment as a source of learning to the traditional culture of society so expected learning process able to give birth to intellectual students and appreciate local cultures. Therefore, it is necessary to arrange learning tools that contain these components so that it can be used to direct the learning process that will take place in the

Science is an approach to studying nature (McLelland, 2003). Science as an academic discipline involves learning concepts as well as processes (Guevara &Almario, 2015). Physics subjects become very important position in society because physics is always around us in everyday life. Physics is not taught simply by providing an understanding of the notions, facts, concepts, principles but also discovery through the process of searching by real action. Based on the characteristics of physics, physics learning at this time is not only emphasized on the product but also on the process. Scientific process skills are the thinking skills used to build conceptual knowledge so as to solve problems and formulate results (Ozgelen, 2012). The development of understanding in science depends on the ability to perform process skills in scientific behavior (Harlen, 1999). In line with that, in the process of physics learning itself there are characters that can be implanted to students so that after the learning of the students not only have the skills but also the formation of student characters. One of the characters that can arise from science learning especially physics is curiosity.

Observation of the authors of the observation of learning in the classroom and interviews of teachers and students in SMA N 1 Padang Bolak show the fact in the field that the learning instrumentss used by teachers more dominant using the book package obtained without any modification and adjustment of the teacher so it looks no the process of developing student knowledge. In addition, learning tools used by teachers in the classroom are still less attention to the environment as a source of learning where the local culture in the community and the environment can be used as a learning material that is associated with physics itself so that students have a meaningful science process skills because of observing and build their knowledge through the cultural facts that exist in the vicinity as well as students' understanding of local socio-cultural values that can provide a positive influence on the formation of student characters.

The scientific inquiry learning model, students are guided by the teacher to understand physics. Furthermore, scientific inquiry based learning can improve the science process skills and students' attitudes toward science (Ergul et al. 2011; Turpin 2004). The core of the scientific inquiry learning model involves students in investigating the real problem by confronting them in the investigation, helping them identify methodological or conceptual problems in addressing the problem. The instructional effect of the scientific inquiry learning model is the research process (Joyce, 2009). In addition, the learning process using scientific inquiry model based on Batak culture is expected to help students in the process of physics learning itself. This is in line with the results of research that explains the importance of science curriculum based on local culture can facilitate students in learning by linking aspects of culture in science learning (Hardoyono, 2007).

Stages in scientific inquiry learning is an inquiry activity that can foster students' curiosity. This is in accordance with the opinion of Pluck and Johnson (2011) which states that inquiry learning has an advantage in generating student curiosity that is part of intrinsic motivation. The indicators of curiosity used in this study are according to Harlen (2000) modified with indicators according to Kemendiknas (2010).

Scientific process skills are a process whereby exploring and obtaining the evidence used to develop ideas (Harlen&Elstgeest, 1992). The science process skill indicators used are observing, asking questions, formulating hypotheses, predicting, finding patterns and relationships, communicating effectively, designing and creating, designing and planning inquiry, manipulating materials and tools effectively, and measuring and counting. The purpose of this research is to 1) describe the quality of learning instruments using scientific inquiry model based on Batak culture on static fluid material in terms of validity, practicality, and effectiveness, 2) to know whether the science process skill of students taught by learning instruments of scientific inquiry model based on Batak culture can increase, and 3) to know whether the curiosity of students taught by learning tools using scientific inquiry model based on Batak culture can increase.

2. Method

The population in this study is all students X SMA Negeri 1 Padang Bolak. The sample technique is done by purposive sampling. The subjects of small group development of learning instruments are X class MIA students with 10 students and the subjects of limited class trial are 32 students of X MIA SMA Negeri 1 Padang Bolak.

The variables in this study are: independent variables in this study is scientific inquiry model inquiry based on Batak culture, and the dependent variable in this research is science process skill and student's curiosity.

This research includes the type of development research by using Thiagarajan, Semmel and Semmel development model that is modified 4-D model (define, design, develop, disseminate). Learning tools developed consist of Lesson Plan (LP), Student Work Sheet (SWS), Student Book (SB), Teacher Book (TB).

Research instrument consisting of student science skill test activity, student's curiosity observation sheet, student activity observation sheet, learning management observation sheet, student response sheet, and instructional instruments validation sheet.

The experimental design used in this study is one-group pretest-posttest design that can be shown in Table 1.

Table 1. Research Design of One-Group Pretest-Posttest

Pretest	Treatment	Posttest
O_1	X	O_2

(Sugiyono, 2010)

Note :

O1 : Prettest of Scientific Process Skill

X : Treatment with learning instruments with model scientific inquiry based batak culture

O2 : Posttest Scientific Process Skill

3. Result

Based on the results of the analysis and discussion in this study, presented the results of research learning instrumentss that have been developed and have been tested otherwise valid, prkatis, and effective. This is because the learning tools that have been developed meet the criteria of validity, practicality, and effectiveness based on the criteria:

a. The four validators concluded that the LP, SWS, Student Books and Master Books, student's curiosity observation sheets, and science process skill tests can be used with minor revisions and these revisions have been corrected in accordance with the validator streaks.

b. Student positive response for 32 students reach 90.97% that is included in the category of positive response.

c. The overall average score on the assessment of the implementation of teaching materials component is in the practical criteria of 3.62 including the high category.

d. Student activities in each meeting during the learning process using learning tools based on scientific inquiry model based on Batak culture in the trial of limited classes has increased.

e. The level of completeness of the students' science process skills in the classroom trial is limited to 87.5%. Improving students' science process skills using learning tools based on scientific inquiry model based on Batak culture on static fluid material with normalized gain of 0.66 with medium category.

f. Increasing students' curiosity using learning tools based on scientific inquiry model based on Batak culture on static fluid material is experiencing an increase in some indicators of curiosity viewed from the increase of graph in each meeting that is on the following indicators: 1) enthusiastic to seek answers each questions, 2) attention to the observed object, 3) enthusiastic on the science process, 4) putting on every step of the activity.

4. Discussion

4.1. Instruments Development Product Valid through Scientific Inquiry Model Based on Batak Culture

Learning instruments development products are said to be valid if the assessment of the four validators on the whole learning instruments is in the "quite valid" category. The validation results of the overall learning tools developed are presented in table 2 below:

Table 2. Validation Result of Learning Instruments

Developed Instruments	Category
Lesson Plan	Valid
SWS	Valid
Students Book	Valid
Teacher Book	Valid
Science Process Skill Test	
Observation Sheet of	Valid
Student Curiosity	Valid

4.2. Practical Instruments Development Products through Scientific Inquiry Model Based on Batak Culture

Learning instruments development products are said to be practical if they meet the criteria in terms of: 1) expert assessment / practitioners of developed learning tools can be used with little revision; 2) in terms of the average total component of teaching materials is 3.62 including high category.

4.3. Effective Instruments Development Products through Scientific Inquiry Model Based on Batak Culture

The effectiveness of learning tools developed through scientific inquiry model based on Batak culture has fulfilled the effective criteria in terms of: 1) students 'learning completeness in the classical that is at least 80% of students who follow the learning able to achieve score ≥ 75, where students' learning completeness in classical reach 87 , 5% with learning tools developed; (2) students' positive response reaches 90.97%; (3) student activity in every learning meeting has increased. Based on posttest analysis result of limited class test, it is found that science process skill has fulfilled classical mastery criteria. This is due to the material and the problems that exist in the student book and student activity sheets developed in accordance with the condition of student learning environment and refers to the study of scientific inquiry model based on Batak culture. The learning process that takes place can be understood more easily by the students because the material taught relates to the culture and environment of everyday students so that students can learn concepts in concrete and real. based on the results of research Hardoyono (2007) also obtained that the curriculum should pay attention and care about the socio-cultural system that develops and applies in a society. Similarly, the development of IPA curriculum needs to integrate the traditional science load so that the learning process becomes meaningful and contextual. In addition, the results of the research from Hussain et al. (2011) explained that there is a significant difference between scientific inquiry and combination scientific inquiry method with conventional learning toward student learning result, which is characterized by students' ability in applying physics concepts in real life.

4.4. Improving Student Science Process Skills

Based on data of pretest and posttest score of science process skill Student obtained got minimum score (Xmin), maximum score (Xmax), average value () and gain as follows:

Table 3. RekapitulasiSkorTes KPS

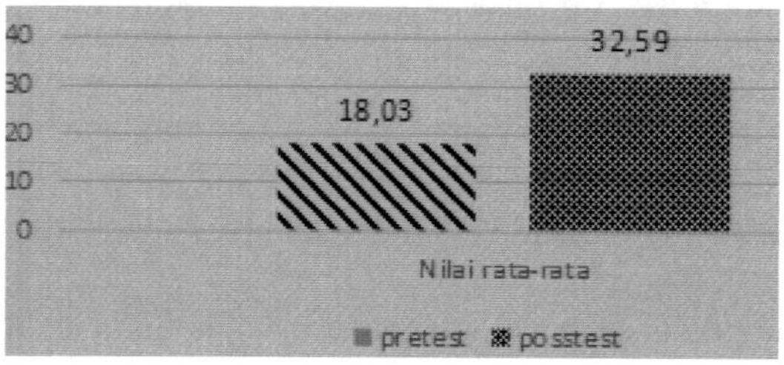

Tes	$\overline{X}$	Gain	<g>	**category**
Pretest Posttest	18,03 32,59	14,56	0,66	Moderate

Figure 1. Improved science process skills on a Limited Classroom Trial.

Based on table 3 and figure 1 it can be seen that after using teaching materials of scientific inquiry model based on Batak culture, the result of students' science skill test is increased which is indicated by posttest average value which is bigger than pretest. Based on the average of pretest and posttest score, obtained normalized gain and gain. The normalized gains can show that in general the skills of the science process have increased by a value of 0.66 and belong to the moderate category. Based on the above description it can be said that the science process skills of students increased in the medium category after learning by using learning tools using scientific inquiry model based on Batak culture.
The application of learning with scientific inquiry model can improve the skill of science process, this is in accordance with research conducted by Nasution (2015) stating that with science inquiry learning can improve students' science process skill. The results of a study by Ergul&Remziye (2011) suggest that inquiry-based learning provides a significant improvement in the students' science skill score in the experimental class. This is supported by Simsek and Kabapinar (2010) stating that inquiry based learning can have a positive influence on students' science process skills. This is possible because students are trained to conduct a scientific investigation, finding the solution of the problem so as to practice their skills.

4.5. Increasing Students' Want to Know

The increase of students' curiosity was obtained by using the observation format of student's curiosity observer during the learning activity. The activities of the students studied include 5 indicators of students' curiosity, enthusiastic searching for answers, attention to observed objects, enthusiastic in the science process, asking every step of the activity, and asking or reading various sources related to static fluid material. Based on observation data of students' curiosity observed during three meetings when applied scientifically speaking inquiry model based on Batak culture, it can be analyzed from 5 indicators of curiosity of students each meeting as follows:

a. Enhusiastic Looking for Answers

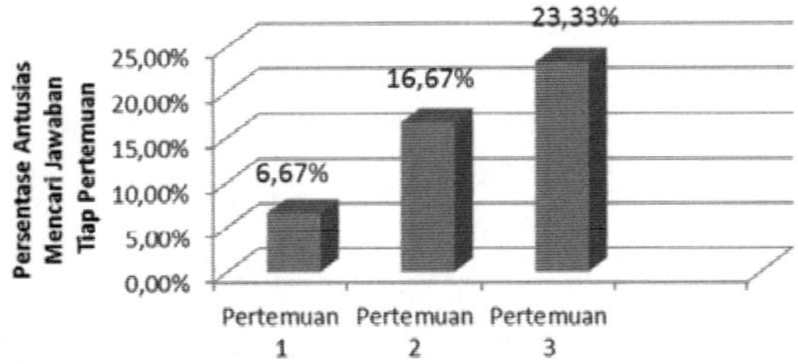

Figure 2. Enthusiastic Percentage Bar Diagram Seeking Answers to Each Meeting.

b. Attention to the Object being Observed

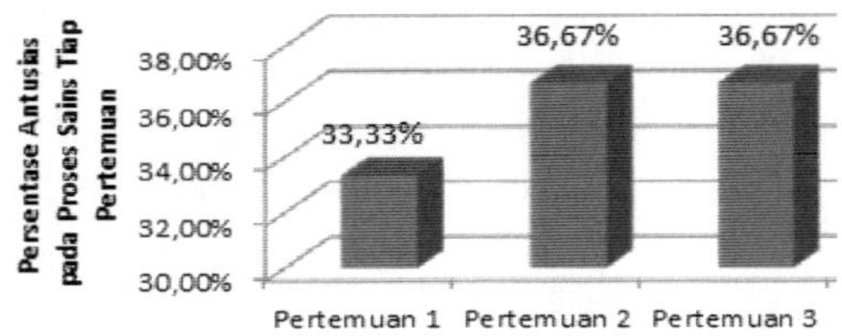

Figure 3. Stage Percentage Attribute Chart of Observed Objects at Each Meeting.

c. Enthusiastic on the Science Process

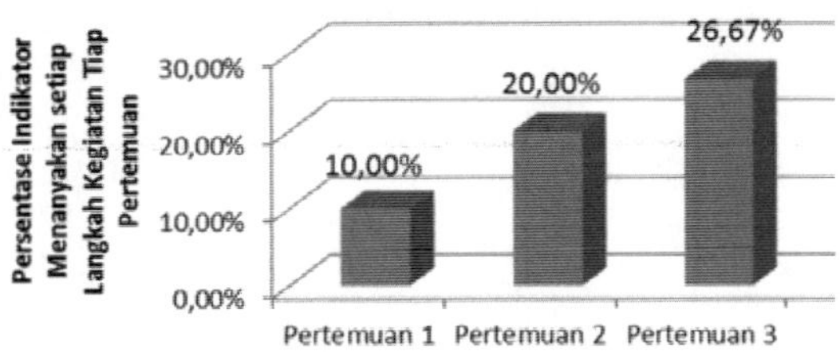

Figure 4. Enthusiastic Bars Chart on the Science Process at Each Meeting.

d. Asking Every Step of Activity

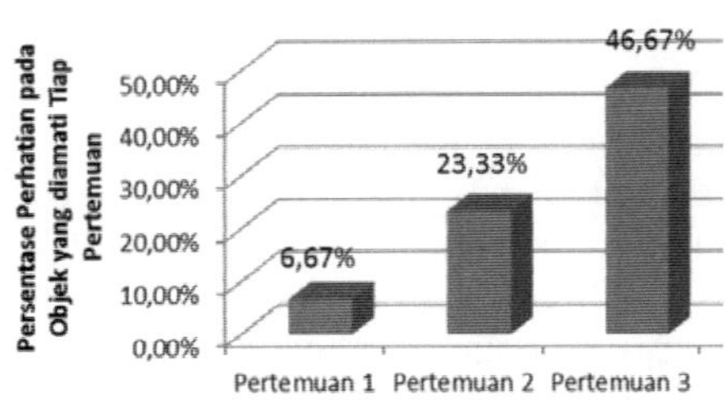

Figure 5. The Bar Diagram Asks Each Activity Step at Each Meeting

e. Questioning or Reading several Relevant Source with Static Fluid

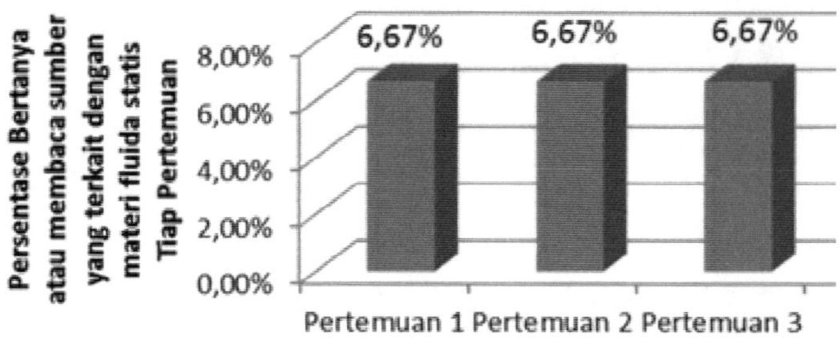

Figure 3. Asking Bar Diagrams or Reading Various Sources Related to Static Fluid Materials at Each Meeting

Based on the five indicators of students 'curiosity, it can be concluded that the application of scientific inquiry model based on Batak culture can increase students' curiosity, although there are some visible indicators that do not experience positive change. However, more curiosity indicators of students are experiencing positive changes from each meeting. students can be concluded that the application of scientific inquiry model based on Batak culture can increase students' curiosity although there are some visible indicators that do not experience positive change. However, more curiosity indicators of students are experiencing positive changes from each meeting.
Fun learning activities can increase students' curiosity. Learning that is integrated with the culture can make students interested in the subject matter. Litman&Spierlberger cited in Mega, et al (2010) stated that students' curiosity can be stimulated by providing visual information. Student curiosity can be improved by providing interesting visual information. Binson (2009) expressed curiosity can foster internal motivation to learn and understand something so that curiosity can be developed in the learning process of science.

5. Conclusion
Based on the results of analysis and discussion in this study, presented several conclusions as follows:

The validity of learning tools developed included in the valid category with the average value of total validity of LP of 3.88 with the category is quite valid, the book students of 4.32 valid categories, teacher books for 4.19 valid categories, and SWS of 4.10 with valid category. The effectiveness of learning tools developed through scientifically speaking inquiry model based on Batak culture has met the effective criteria in terms of: (1) students' learning completeness in the classical that is at least 80% of students who follow the learning able to achieve score ≥ 75, where student learning completeness in classical reached 87.5% with developed learning tools; (2) students' positive response reaches 90.97%; (3) student activity in every learning meeting has increased. The practicality of learning tools developed through scientifically based inquiry model based on Batak culture has met the criteria in terms of: (1) expert assessment / practitioners of developed learning tools can be used with little revision; (2) in terms of the total component of teaching materials is 3.62 including high category. Improving students' science process skills using learning tools based on scientific inquiry model based on Batak culture on static fluid material with normalized gain of 0.66 with medium category. Increasing the curiosity of students using learning tools of scientific inquiry model based on Batak culture on static fluid material is experiencing an increase in some curiosity indicator that is viewed from the increase of graph in every meeting, that is on the following indicators: (1) enthusiastic looking answer each question, (2) attention to the object being observed, (3) enthusiastic on the science process, (4) ask every step of the activity.

References

[1] Binson, B. 2009. *Curiosity Based Learning (CBL) Program*. US-China Education Review

[2] Danoebroto, S. W. 2012. Model Pembelajaran Matematika Berbasis Pendidikan Multikultural. *Jurnal Pembangunan Pendidikan : Fondasi dan Aplikasi*, Vol 1, No 1, 94-107

[3] Depag RI. 2006. Undang-Undang Republik Indonesia No.20 Tahun 2003; Sistem Pendidikan Nasional. Jakarta : Sekretaris Ditjen Pendidikan Islam.

[4] Dhakaa, A. 2012. Biologycal Science Inquiry Model And Biology Teaching. *Bookman International Journal of Accounts, Economic & Business Management*, Vol 1 No 2. (http://www.bookmanjournals.com, diakses 14 Oktober 2016).

[5] Ergul & Remziye. 2011. The Effects of Inquiry-Based Science Teaching On Elementary School Students' Science Process Skill and Science Attitudes. *Bulgarian Journal of Science and Education Plicy (BJSEP),* (online),Vol. 5 No. 1. (http://bjsep.org, diakses 21 September 2016)

[6] Guevara & Almario, C. 2015. Science Process Skills Development Through Innovations in Science Teaching. *Research Journal of Educational Sciences*, 3(2): 6—10.

[7] Harlen, W. 1999. Purposes and Procedures for Assessing Science Process Skills. *Assesment in Education: Principles, Policy & Practice*, 6(1), 129—144

[8] Harlen, W. & Elsgesst, J. 1992. *UNESCO Sourcebook for Science in the Primary School*. France: Imprimerie de la Manutention.

[9] Harlen, W. 2000. *Teaching, learning and assessing science 5-12*. London Paul Chapman Publishing Ltd.

[10] Hake, R. R. 1999. Analyzing Change/Gain Scores. *American Educational Research Association's Division:* 1-4.

[11] Hardoyono, F. 2007. Tinjauan Aspek Budaya pada Pembelajaran IPA: Pentingnya Pengembangan Kurikulum IPA Berbasis Budaya Lokal. *Jurnal Pemikiran Alternatif Pendidikan.* Vol 12. No.2, 143-163.

[12] McLelland, C.V. 2003. The Nature of Science and The Scientific Method. (Online), (http://www.geosociety.org/educate/NatureScience , diakses 12 Oktober 2016)

[13] Mega, P,. Fakhruddin, Z,. Maa'ruf, Z,. 2010. Implementation of Conceptual Understanding Procedures to Improve Students' Curiosity in Learning Physics at Class VII SMPN 4 Pekanbaru. *Jurnal Online UNRI*

[14] Nadelson,L. Williams,S. Turner,H. 2008. Influence of Inquiry-Based Science Intervention on Middle School Students' Cognitive, Behavioral, and Affective Outcomes

[15] Nieveen, N. & Plomp, T. 2007. *An Introductionto Educational Design Research*. Enschede. Netzodruk

[16] National Research Council. 2000. *Inquiry and the National Science Education Standards*: A guide for teaching and learning. Washington DC: National Academy Press

[17] Nasution, P. 2005. *Adat Budaya Mandailing dalam Tantangan Zaman*. FORKALA Prov.Sum.Utara

[18] Özgelen, S. 2012. Students' Science Process Skills within a Cognitive Domain Framework. *Eurasia Journal of Mathematics, Science & Technology Education,* 8(4): 283—292.

[19] Philips, R. 2014. Space for Curiosity. *Journal of Progress in Human Geography*. 38(4) 493-512.

[20] Pluck, G & Johnson, H. 2011. *Stimulating Curiosity to Enhance Learning : Education Science and Psychology*. No.2(19).

[21] Sugiyono. 2010. *Metode Penelitian Kuantitatif, Kualitatif, dan R&D*. Bandung : CV. Alfabeta

[22] Simsek, P., & Kabapinar, F. 2010. The Effects of Inquiry Bases Learning on Elementary Students' Conceptual Understanding of Matter, Scientific Process Skills and Science Attitudes. *Procedia Social and Behavioral Sciences Journal*, (online), (2) 1190-1194. (www. Sciencedirect.com, diakses 28 Agustus 2016)

.

The Effort to Reduce a Muscle Fatigue Through Gymnastics Relaxation and Ergonomic Approach for Computer Users in Central Building State University of Medan

Syamsul Gultom[1], Indra Darma Sitepu[1] and Nurman Hasibuan[1]

[1]Faculty of Sport Science, Universitas Negeri Medan

Corresponding author's : syamsulgultom@gmail.com

Abstract. Fatigue due to long and continuous computer usage can lead to problems of dominant fatigue associated with decreased performance and work motivation. Specific targets in the first phase have been achieved in this research such as: (1) Identified complaints on workers using computers, using the Bourdon Wiersma test kit. (2) Finding the right relaxation & work posture draft for a solution to reduce muscle fatigue in computer-based workers. The type of research used in this study is research and development method which aims to produce the products or refine existing products. The final product is a prototype of back-holder, monitoring filter and arranging a relaxation exercise as well as the manual book how to do this while in front of the computer to lower the fatigue level for computer users in Unimed's Administration Center. In the first phase, observations and interviews have been conducted and identified the level of fatigue on the employees of computer users at Uniemd's Administration Center using Bourdon Wiersma test and has obtained the following results: (1) The average velocity time of respondents in BAUK, BAAK and BAPSI after working with the value of interpretation of the speed obtained value of 8.4, WS 13 was in a good enough category, (2) The average of accuracy of respondents in BAUK, in BAAK and in BAPSI after working with interpretation value accuracy obtained Value of 5.5, WS 8 was in doubt-category. This result shows that computer users experienced a significant tiredness at the Unimed Administration Center, (3) the consistency of the average of the result in measuring tiredness level on computer users in Unimed's Administration Center after working with values in consistency of interpretation obtained Value of 5.5 with WS 8 was put in a doubt-category, which means computer user in The Unimed Administration Center suffered an extreme fatigue. In phase II, based on the results of the first phase in this research, the researcher offers solutions such as the prototype of Back-Holder, monitoring filter, and design a proper relaxation exercise to reduce the fatigue level. Furthermore, in in order to maximize the exercise itself, a manual book will be given to employees whom regularly work in front of computers at Unimed's Administration Center

1. Introduction

Occupational health is a specialization in health sciences that aims to improve the highest degree of health both physically and mentally with prenventif and curative efforts against diseases caused by occupational factors and work environment. Act No. 13 of 2003 article 86, paragraph 2 states that the work safety and health effort is intended to provide safety assurance and improve the health status of the workers by preventing accidents and occupational diseases, hazards in the workplace, health promotion, treatment and rehabilitation.

According to Depkes RI (2005), 40.5% of workers in Indonesia have health problems related to their work and among them are skeletal muscle members as much as 16% (Tana 2009). Based on data in Indonesia General Hospital of Lampug Tengah, a group of musculoskeletal diseases including back pain was ranked the top ten most diseases in outpatients. During the year 2006 there were 32 patients treated with muscle pain complaints, whereas in 2007 there was a surge more than triple. Most of them suffer from rheumatoid arthritis and some back pain. During January to March 2008 there were about 40 patients with the same pain complaints and most of the sufferers were workers who spent much time in office as computer operators (Sumekar, 2008).

The total ergonomic approach emphasizes the use of a systemic, holistic, interdisciplinary and participatory approach (SHIP approach) in the analysis of ergonomic issues and formulates workplans and selects interventions through the application of appropriate technology (TTG). Ergonomic interventions are conducted thoroughly resulting in best interventions with minimal impact (Manuaba, 2006). The implementation of the total ergonomic approach is recommended by some experts because the method used is able to create humane, competitive and sustainable working conditions thereby increasing the productivity and income of employees and employers (Manuaba, 2005).

Relaxation is one of self-management techniques based on the workings of the sympathetic and parasympathetic nervous system. In Indonesia, research on relaxation is also quite a lot done. Prawitasari (1988), reported that relaxation is useful for reducing physical fatigue.

Based on the initial survey conducted on the employees of computer users in the Central Administration of Medan State University obtained information that they most experienced complaints in the neck, shoulder and waist. Fatigue of these muscles also resulted in their performance decreased so that data should always be updated often experience delays and not accumulated well, and often occur data revision because some data have typing errors. The operator works from 08.00 - 16.00 WIB and is given one time break which is 12.00-13.00 WIB. At the time of obtaining tasks such as data processing KRS and KHS, charging DPNA, and even processing data payments SPP student money, then the time to rest will be shorter and the end of the work will be longer than the time specified. From the data above shows that there is an effort to decrease the level of muscle fatigue through relaxation exercises and ergonomics approach for employees of computer users in Unimed Administration Center.

2. Research Methods

The method used in this research is research development with the method "Research and Development" which aims to produce products or improve existing products that can be justified. The resulting product is a backstop design, monitor filters and relaxation gymnastics design and the provision of a pocket book design how to do relaxation exercises while in front of the computer to reduce fatigue levels for computer user employees at the Unimed Administration Center. The Borg & Gall development model includes a systematic guide of steps undertaken by a researcher in order for the product to be designed to have a feasibility standard in the field. Borg & Gall (2007: 590) states that the development research procedure basically consists of two main objectives: (1) developing the product as a function of the developer, and (2) testing the effectiveness of the product as validation in achieving the goal. Thus, the concept of development research is more properly defined as a development effort which is also accompanied by validation efforts. The population of this research is civil servant or honorary staff who use computer in building Administration Center Unimed amounted to 67 people that is computer user in Academic Administration and Students (BAAK) Bureau counted 24 Respondent, Administrative Bureau of General and Finance (BAUK) 32 Respondents and Administrative Bureau Planning and Information Systems (BAPSI) 11 Respondents who use computers over 4 hours 1 day.

IOP Conf. Series: Journal of Physics: Conf. Series **970** (2018) 012010 doi:10.1088/1742-6596/970/1/012010

3. Results

In this study of muscle relaxation techniques that have been validated by the expert and back buffer design, monitor filters to reduce fatigue rates for computer users in the Center Administration Unimed as follows:

3.1. Technique of muscle relaxation movement that has been validated by the expert

3.1.1. Relaxation of the waist muscles

1. Movement rotate waist to right and left with count 1 to 8
2. Hold both hands and do a light beating at the waist
3. Bend down and straighten up

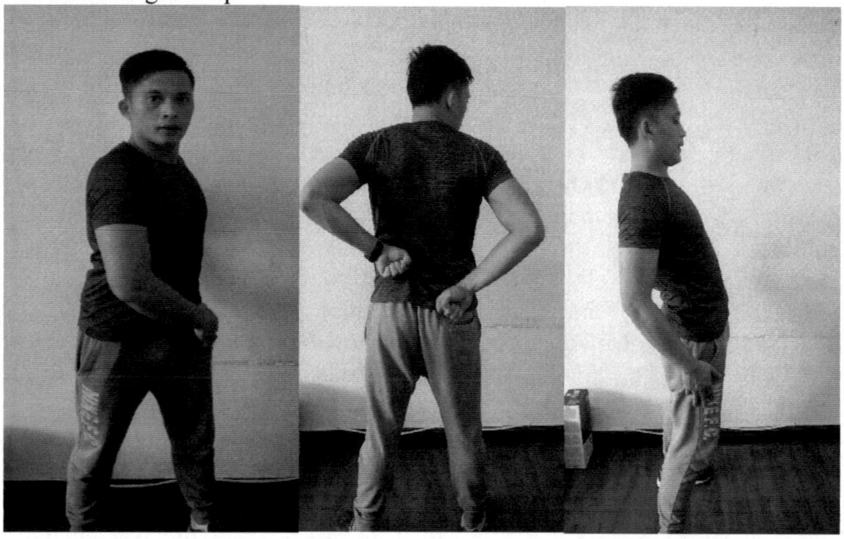

Figure 1. Waist Muscles Relaxation Movement

3.1.2. Relaxation of the back muscles

1. Straighten both hands forward and pull backward with interpal hands in the fore gear while exhaling from the mouth, and both hands are pulled backward while drawing breath from the nose.
2. Rotate both hands with a straight elbow position in the pull backward from front to back and vice versa the back.
3. Align the right hand from the front in the drag sideways 1 to 8 and replaced with the left hand with the same movement.

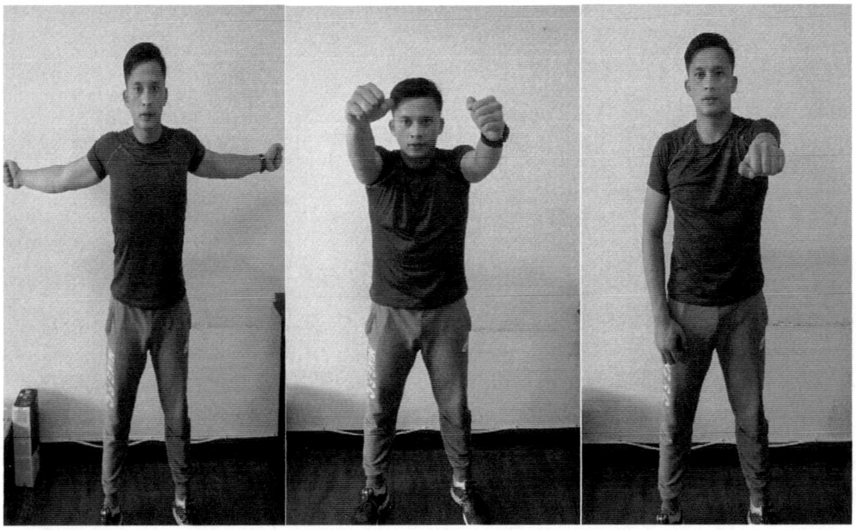

Figure 2. Back Muscle Relaxation Movement

3.1.3.. Relaxing the hand muscles and fingers

1. Sticking the fingers of both hands and make movements such as the waves with the elbows bent hands in front of the chest.
2. Gluing the two fingers with the palm of the hand outward with the elbow position in straightening in alternating bend.
3. Gluing both fingers with palm position into the inside. both elbows straight. The hands alternately pull to the left and right side alternately.

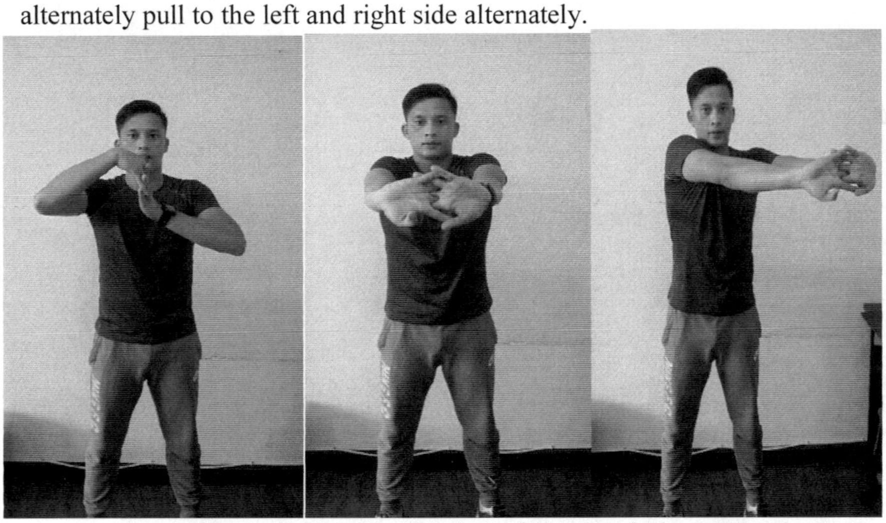

Figure 3. Relaxing Motion of Hand and Finger Muscles

3.1.4. Relaxation of the neck muscles

1. Rotate the left half circle neck and to the right
2. Bend the neck to the right and left side
3. Bend the neck forward and backward.

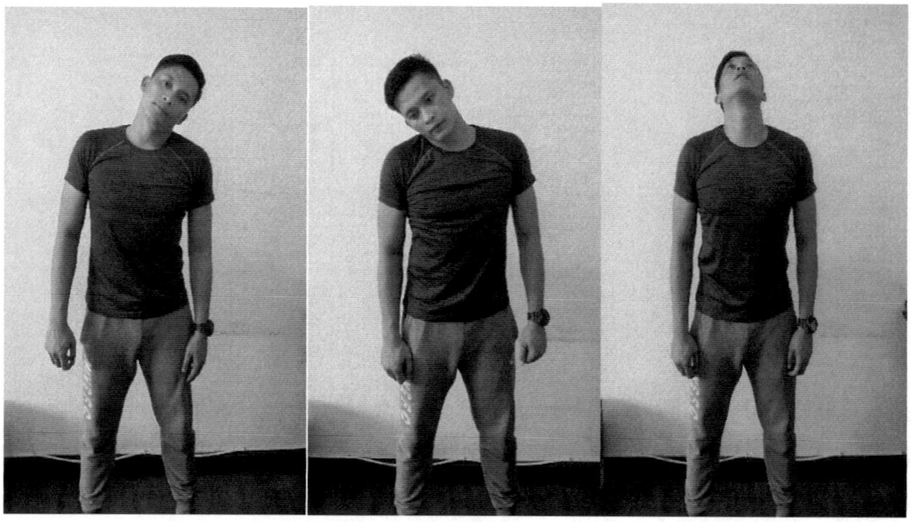

Figure 4. Neck Muscle Relaxation Movement

3.1.5. Eye muscle relaxation

1. Closes and opens eyes with a count of one close and count 2 opens up to 8 counts
2. Pressing with both fingers at the bottom of the eyelid
3. Pressing with your finger over the eyelid.

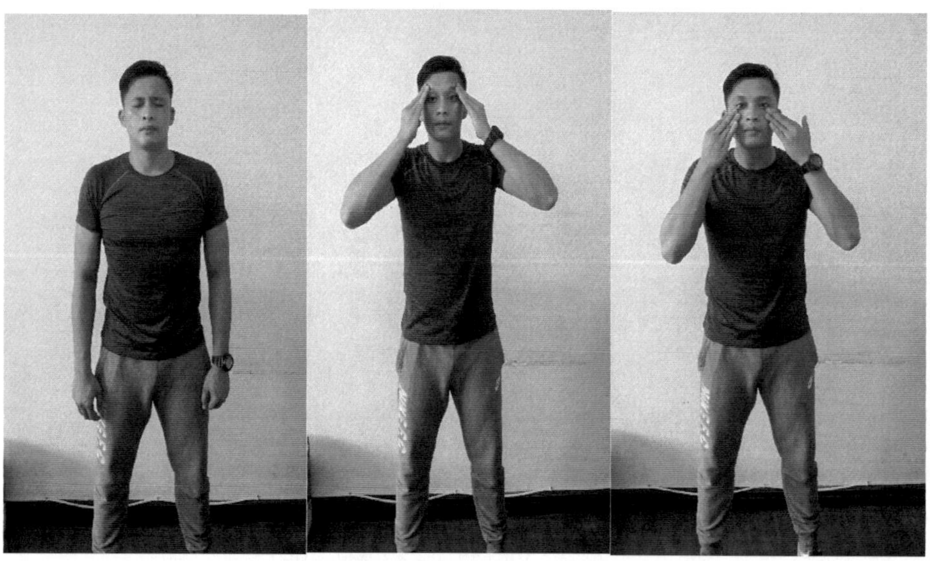

Figure 5. Eye Muscle Relaxation Movement

3.2. Back-Holder design and Monitoring Filter

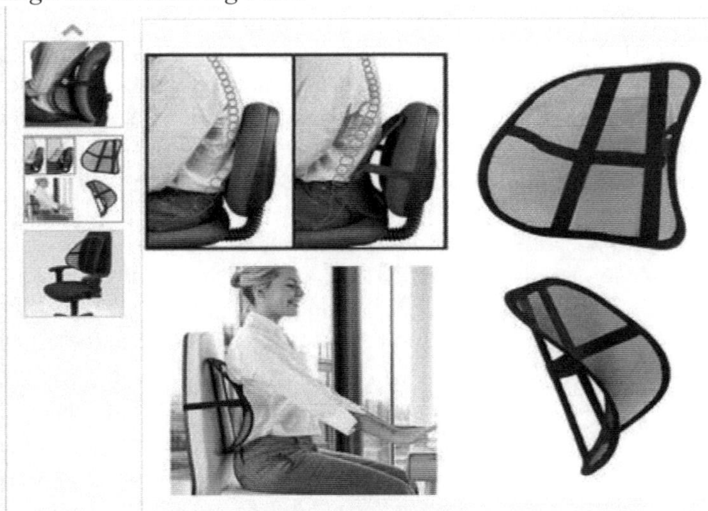

Figure 6. Back-Holder

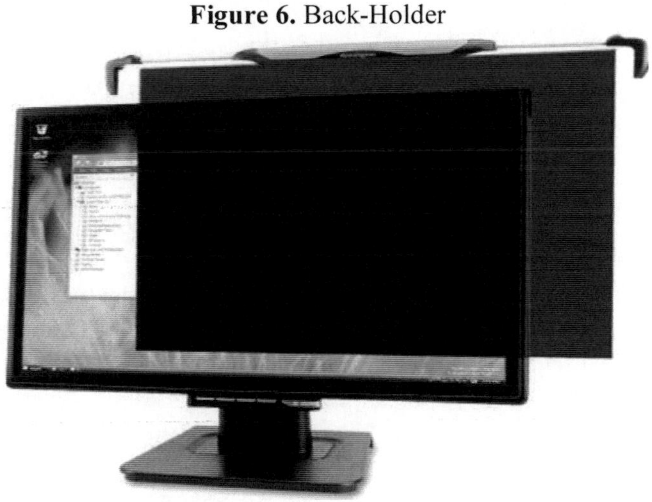

Figure 7. Monitoring Filter

4. Conclusion

Based on this finished study, it can be concluded that the result of the first phase clearly shown the holistic tiredness in computer users at Unimed's Administration Center. The general description regarding to the health issue which suffered by the employees can at least be identified as a following action. The researcher devises solutions like a prototype of Back-Holder, Monitoring Filter, dan a Manual Book of how to do the exercise which has been validated by experts in order to decline the level of fatigue when they are using computers at Unimed's Administration Center. Moreover, this research could possibly assist the computer users to cope with their health issue.

References

[1] Borg, Walter R. & Meredith G, *Educational Research An Introduction*. New York: Longman, 2007.

[2] Departemen Kesehatan R.I. (2005). Rencana Strategi Departemen Kesehatan. Jakarta: Depkes RI.

[3] Manuaba, A., 2005. Pendekatan Holistik dalam Aplikasi Ergonomi. Sosial & Humaniora. Okt;01(01):1-13.
[4] Manuaba, A. 2006. Macro ergonomics approach on work organizations with special reference to the utilization of Total ergonomic SHIP approach to obtain humane, competitive and sustainable work system and products. Proceeding Seminar Nasional Ergonomi. Surabaya, 21-22 November 2006.
[5] Prawitasari. J.E. (1988), 'Pengaruh Relaksasi terhadap Keluhan Fisik'. Laporan Penelitian.Yogyakarta: Fakultas Psikologi UGM
[6] Sumekar, DW., Deny, N. 2008. Nyeri Punggung pada Operator Komputer Akibat Posisi dan Lama Duduk. Diakses tanggal 19 Mei 2015
[7] Tana, L., Delima, D., dan Tuminah, S., 2009. Hubungan Lama Kerja dan Posisi Kerja dengan Keluhan Otot Rangka Leher dan Ekremitas atas pada Pekerja Garmen Perempuan di Jakarta Utara. Diakses tanggal 23 Februari 2015

IOP Conf. Series: Journal of Physics: Conf. Series **970** (2018) 012011 doi:10.1088/1742-6596/970/1/012011

Flood Discharge Analysis with Nakayasu Method Using Combination of HEC-RAS Method on Deli River in Medan City

Rumilla Harahap[1*], Kemala Jeumpa[1], Bambang Hadibroto[1]

[1]Department of Civil Engineering, State University Of Medan, Indonesia

*Corresponding Author: rumi_harahap@yahoo.com

Abstract - The problem in this research is how in the rainy season the water does not overflow, does not occur flood and during the dry season does not occur drought so it can adjust the condition or existence of Deli river which is around Medan city. Deli River floods often occur, either caused by a smaller capacity than the existing discharge, lack of maintenance and drainage and disposal systems that do not fit with the environment, resulting in flood subscriptions every year. The purpose of this research is to know flood discharge at Deli river as Flood control in Medan city. This research is analyzed on several methods such as log Pearson, Gumbel and hydrograph unit, while HEC-RAS method is modeling conducted in analyzing the water profile of the Deli River. Furthermore, the calculation of the periodic flood discharge using the Nakayasu Method. Calculation result at Deli River return period flood discharge 2 years with an area of 14.8 km^2 annual flood hydrograph the total is 26.79 m^3/sec on the hours at the 4th time . Return period flood discharge 5 years with an area of 14.8 km^2 annual flood hydrograph the total is 73,44 m^3/sec. While 25 annual return period total flood hydrograph is 146.50 m^3/sec. With flood analysis can reduce and minimize the risk of losses and land can be mapped if in the area there is flooding.

1 Introduction

At this time is often the occurrence of floods in every area where the flood is an event where there is excessive water overflow somewhere. According to [1] The high rainfall and the inability of the river to accommodate the rain water is blamed as the culprit of the flood. Thousands of people became victims, homes of residents submerged, even some residents who died was washed away by floods. The presence of uniform rainfall, high intensity and continuous in some areas in the city of Medan has caused many floods. So many residents are displaced in a safe place [2].

Hydrologically watershed management is an attempt to manage the Earth's surface biophysical conditions in such a way as to obtain a maximum water discharge yield and have a distributed flow regime evenly distributed throughout the year [3]. Based on the hydrological data can be determined the flood index in a river that will be useful in managing the watershed environment. Until now the flood index in the Asahan River based on hydrological data for several periods of flood discharge has never been determined. From the hydrological data obtained an average discharge of 400 m^3/s in the downstream area [4]. The design of flood discharge calculations were performed using the Nakayasu Hydrograph Synthetic Unit method. The Nakayasu Hidrograph Synthetic Unit is widely used in dam planning and river improvement in Brantas (East Java) projects, among others to determine the discharge planning of Lahor Dam, Sengguruh, Wonorejo and Brantas River Basin Improvement

To solve the river problem is closely related to the hydrological aspect, especially the rain problem as a source of water that will be streamed to the river system and runoff as a result does not have a river system to drain into the final disposal at sea. Hydrological design is necessary to know the flow of drainage. [5]. In the River channel planning the return period used depends on the channel function and the rain catchment area to be dried. According to experience, the use of return period for planning: Kwarter Channel (1 year return period), Tertiary Channel (2 year return period), Secondary Channel (5 year return period), Primary Channel (10 year return period) [6]. In the science of statistics is known several kinds of frequency distributions that are widely used in the field of hydrology. The following four types of frequency distributions are most widely used in the field of hydrology such as: Normal Distribution, Normal Log Distribution, Log Person Distribution III and Gumbel Distribution [7].

2 Materials and Methods

2.1 Metode Log Pearson type III

To estimate the magnitude of the flood discharge with a particular period, first the rain data is brought closer to a distribution spread, in order to estimate the magnitude of the flood discharge so not to go far from the reality of the flood that occurred. The most widely used distribution size is the standard deviation. If the spread is very large to the average value then the value of standard deviation will be large, but if the data distribution is very small to the average value then the value of standard deviation will be small. The formula used in the Log Pearson type III method is:

$$\mathrm{Log}\, X_T = \overline{\log X} + G.s \tag{1}$$

1. Mean

$$\overline{\mathrm{Log}\, X} = \frac{\sum_{i=1}^{n} \mathrm{Log}\, X_i}{n} \tag{2}$$

2. Skewness

$$Cs = \frac{n.\sum_{i=1}^{n} (\mathrm{Log}\, X_i - \overline{\mathrm{Log}\, X})^3}{(n-1).(n-2).S^3} \tag{3}$$

3. Standatd Deviation

$$S = \sqrt{\frac{\sum_{i=1}^{n} (\mathrm{Log}\, X_i - \overline{\mathrm{Log}\, X})^2}{n-1}} \tag{4}$$

Probability density function of this distribution are::

$$P'(X) = P_0{}'(X)(1 + X/a)^c e^{-cx/a} \tag{5}$$

2.2 Hydraulic Analysis

The open channels flow consists of natural channels, such as upstream small rivers (mountain) to large rivers in the estuaries, and artificial channels, such as river channels, irrigation channels to irrigate rice fields, sewers, drains to carry water to hydroelectric power, drainage water supply, and flood channels. Artificial channels can be triangular, trapezoidal, rectangular, round, semi-circular, and arranged shapes [6].

Changes in the function of the upstream area of the watershed ± 15% resulted in the balance of the river begin to be disrupted. This disorder contributes to the sharp increase in the quantity of flow discharge and the quantity of sedimentation in the river. It can also be interpreted that an unspoiled river basin and dense vegetation can be changed by 15% of the area without having to change the natural state of the river in question. Land use change can also increase peak discharge 5 to 35 times because the water absorbed into the soil slightly causes the runoff flow to be large, resulting in a large discharge. As shown in Figure 1 and Figure 2. The condition of the river banks in areas before and after develop.

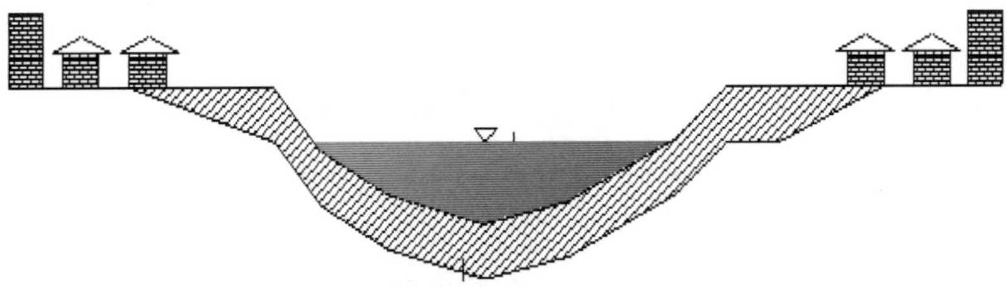

Figure 1. The Condition of River Banks in the Area Before Developing

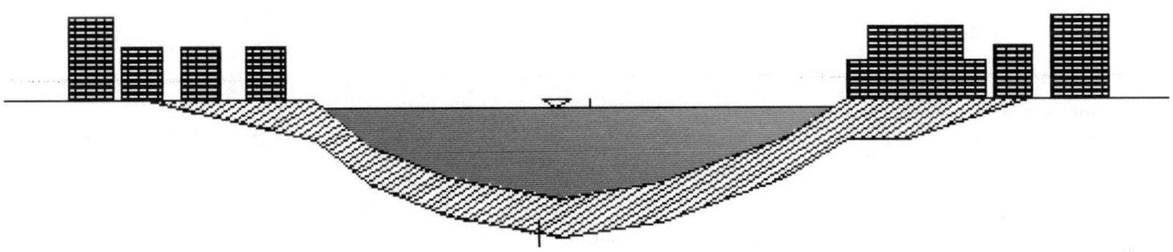

Figure 2. The Condition of River Banks in the Area After Developing

2.3 *General Flow Formulas*

As for finding the flow rate it can be used some Chezy formula (for uniformly fixed flow) $\mathbf{V=C\ (RI)^{0,5}}$

2.4 *Nakayasu Hydrograph Synthetic Unit*

The calculation of the flood discharge design is using the Nakayasu method. The general equation of the Nakayasu Hydrograph Synthetic Unit is as follows:

$$Qp = \frac{C.A.R_0}{3{,}6\,(0{,}3\,Tp + T_{0,3})} \qquad (6)$$

$Tp = tg + 0{,}8\ tr$

$tg = 0{,}21 \times L^{0,7}$ (L < 15 km)

$tg = 0{,}4 + 0{,}058 \times L$ (L > 15 km)

$T_{0,3} = \alpha \times tg$

Where :

Qp	= flood peak discharge (m^3/s)
C	= flow coefficient
R^0	= rain unit (mm)
A	= watershed area (km^2)
Tp	= the grace period from the beginning of the rain to the peak of the flood (hour)
$T_{0,3}$	= time required by the decrease of discharge, from peak discharge to 30% of peak discharge
tg	= concentration time (hr)
tr	= rain time unit, taken 1 hour
α	= hydrograph parameter, worth between 1.5 - 3.5
L	= river length (m)

The research flow in this research can be described as follows:

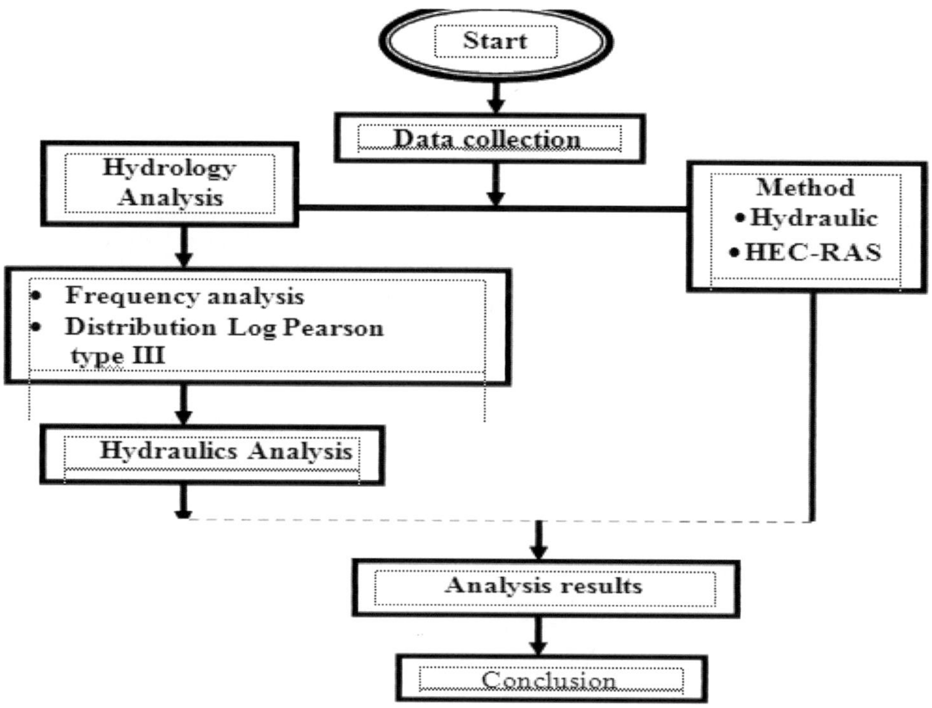

Figure 3. Flow Chart of Research

3 Results and Discussion

Log-Pearson Type III distribution is widely used in hydrological analysis, especially in the analysis of maximum flood discharge and minimum discharge data with extreme values. Of the various types of distribution developed by Pearson, only Type III is the most widely used. The statistical data from Log-Pearson Type III distribution is not close to any of the distribution features. The probability line of Log-Pearson Type III distribution is a curved line, so its usage for flood analysis is often used for its data logarithm (not its own data), so this distribution is called the Log-Pearson Type III distribution.

Table 1. Rainfall Design Return Period Log Pearson Type III Method

Return Period	log X	Cs	G	log X	X (mm)
2	1,9873	0,003353	-0,001	1,987	97
5	1,9873	0,003353	0,842	2,121	132
10	1,9873	0,003353	1,283	2,191	155
25	1,9873	0,003353	1,753	2,265	184
50	1,9873	0,003353	2,057	2,313	206
100	1,9873	0,003353	2,330	2,356	227
1000	1,9873	0,003353	3,098	2,478	301

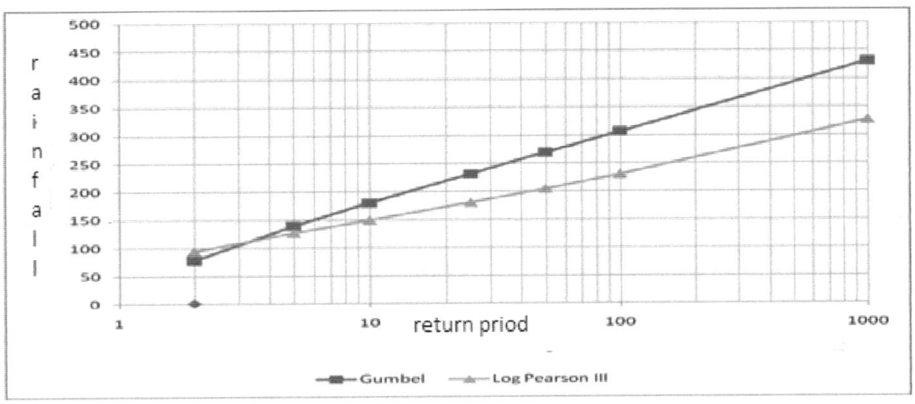

Figure 4. Recapitulation of Pearson III & Gumbell Log Distribution Results

3.1 *Deli Watershed*

From the physical characteristics of the Deli River can be calculated two important elements that will determine the shape of the unit hydrograph, that is Time Lag (TL), Peak time (Tp) and base time (Tb). Height and duration of unit rain that Commonly used is 1 unit of rain duration is generally taken Tr = 1 hour, but can be selected another duration as long as it is expressed in hours then the height of rain every hour should be divided by 2 and distributed in 0.5 hour intervals as seen on Table 2.

Table 2. The Results of the Calculation of Flood Discharge for 2 Year Return Period A= 14,8 km² L= 7 km

t (h)	U (t,1) NAKAYASU (m^3/s/mm)	R1 = 6,45 mm/h	R2 = 7,44 mm/h	R3 = 8,41 mm/h	R4 = 5,03 mm/h	R5 = 1,15 mm/h	Flood Hydrograph (m^3/s)
0	0,000	0,000					0,000
1	0,859	5,541	0,000				5,541
2	1,648	10,631	6,394	0,000			17,025
3	0,840	5,421	12,267	7,219	0,000		24,907
4	0,367	2,365	6,255	13,851	4,316	0,000	26,786
5	0,191	1,231	2,729	7,063	8,280	0,986	20,289

t (h)	U (t,1) NAKAYASU (m^3/s/mm)	R1 = 6,45 mm/h	R2 = 7,44 mm/h	R3 = 8,41 mm/h	R4 = 5,03 mm/h	R5 = 1,15 mm/h	Flood Hydrograph (m^3/s)
6	0,099	0,641	1,421	3,081	4,222	1,891	11,257
7	0,052	0,334	0,740	1,604	1,842	0,964	5,485
8	0,027	0,174	0,385	0,835	0,959	0,219	2,573
9	0,014	0,091	0,201	0,435	0,499	0,114	1,340
10	0,007	0,047	0,104	0,227	0,260	0,059	0,698
11	0,004	0,025	0,054	0,118	0,135	0,031	0,363
12	0,002	0,013	0,028	0,061	0,071	0,016	0,189
13	0,001	0,007	0,015	0,032	0,037	0,008	0,098
14	0,001	0,003	0,008	0,017	0,019	0,004	0,051
15	0,000	0,002	0,004	0,009	0,010	0,002	0,027
16	0,000	0,001	0,002	0,005	0,005	0,001	0,014
17	0,000	0,000	0,001	0,002	0,003	0,001	0,007
18	0,000	0,000	0,001	0,001	0,001	0,000	0,004

From Table 2, the calculation of flood discharge of 2 year return period design A = 14.8 km², L = 7 km maximum flood discharge at 5 to 26,786 m^3/s while starting at 2.00 hrs 17.02 m^3/s . With such a high increase it is necessary to anticipate the discharge by normalizing the Deli river.

Table 3. Flood Discharge Calculation Result of 25 Years Return Period A= 14,8 km² L= 7 km

t (h)	U (t,1) NAKAYASU (m^3/s/mm)	Rl = 35,29 mm/h	R2 = 40,72 mm/h	R3 = 45,98 mm/h	R4 = 27,48 mm/h	R5 = 6,28 mm/h	Flood Hydrograph (m^3/st)
0	0,000	0,000					0,000
1	0,859	30,305	0,000				30,305
2	1,648	58,142	34,967	0,000			93,109
3	0,840	29,646	67,087	39,484	0,000		136,217
4	0,367	12,935	34,207	75,753	23,603	0,000	146,497
5	0,191	6,735	14,925	38,626	45,284	5,391	110,960
6	0,099	3,507	7,771	16,853	23,090	10,343	61,563
7	0,052	1,826	4,046	8,775	10,074	5,274	29,995
8	0,027	0,951	2,107	4,569	5,246	2,301	15,173
9	0,014	0,495	1,097	2,379	2,731	1,198	7,901
10	0,007	0,258	0,571	1,239	1,422	0,624	4,114
11	0,004	0,134	0,297	0,645	0,741	0,325	2,142
12	0,002	0,070	0,155	0,336	0,386	0,169	1,115
13	0,001	0,036	0,081	0,175	0,201	0,088	0,581
14	0,001	0,019	0,042	0,091	0,105	0,046	0,302
15	0,000	0,010	0,022	0,047	0,054	0,024	0,157
16	0,000	0,005	0,011	0,025	0,028	0,012	0,082
17	0,000	0,003	0,006	0,013	0,015	0,006	0,043
18	0,000	0,001	0,003	0,007	0,008	0,003	0,022

3.2 Hydraulics Analysis

The hydraulic analysis of the Deli River cross section on Maimun road is calculated using the HEC-RAS program. With this analysis it can be seen the water level elevation at the river cross section when a water discharge through the river. The data required in river cross section analysis with the help of HEC-RAS software are:

a. Cross-section of the river
b. Cross section of the river
c. Data flow through the river
d. Figures manning cross-section of the river

Before beginning this hydraulic analysis, the necessary data must be prepared. The steps of hydraulic analysis with HEC-RAS program as in Figure 5.

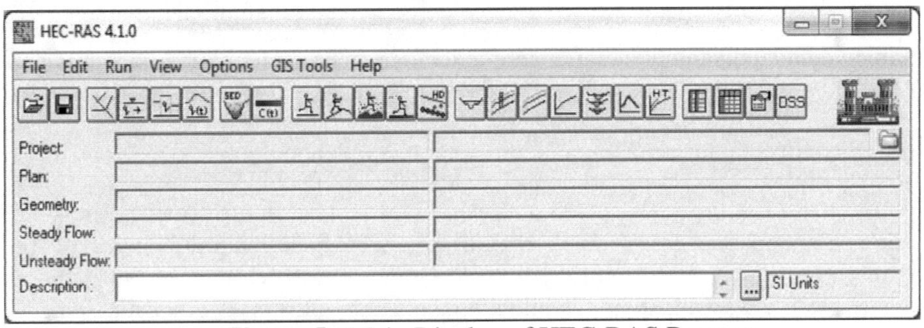

Figure 5 : Main Display of HEC-RAS Program

Draw a river flow by clicking on River Reach. To be able to draw according to the original map, you can use map image file for background drawing by click add/edit background picture. In drawing the river flow the first point made is a river upstream like in Figure 6.

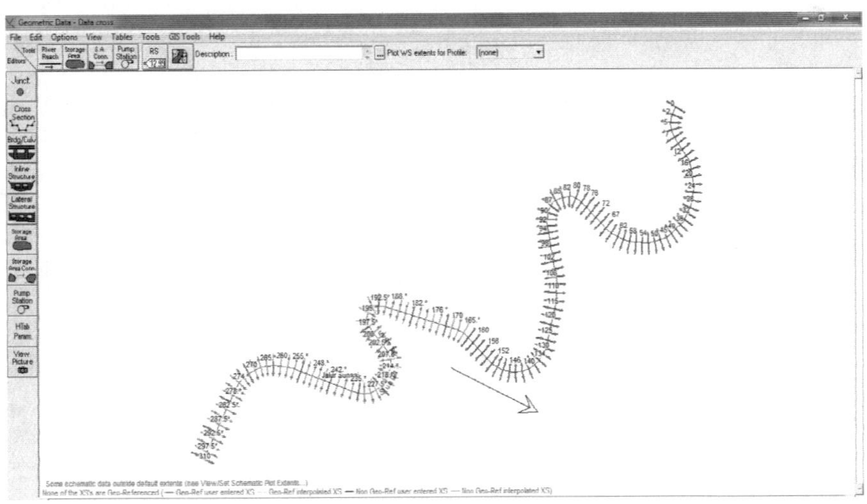

Figure 6 : The Input View of the River Geometry Data of the HEC-RAS Program

Input the cross section then click on the cross section, exit the view as Figure 7. Water profile of the River Deli

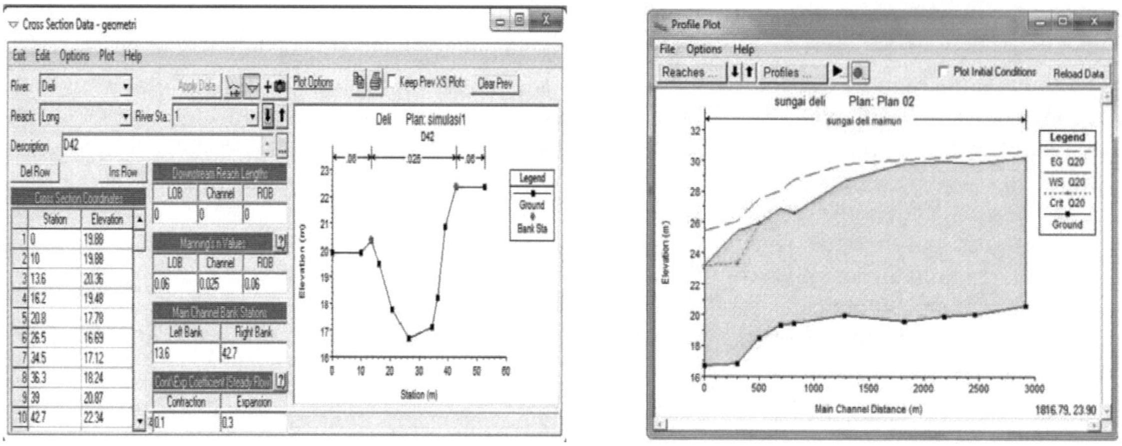

Figure 7: The input interface of the crop data The water profile of the Deli River

3.3 Normalization of Rivers and Levees

The plan to improve the flow is to make a narrow river section improvement. The shape of the river cross section is planned trapezoidal. In determining the cross-sectional dimension of the river it should be noted that existing river morphology is to keep the low flow of discharge remaining. For river bottom elevation still use the existing elevation, so the slope of the river does not change.

Normalization is done by enlarging the cross-sectional dimension of the river on the part of the river where the runoff of the river cross section is made so that no runoff occurs. Section planning is done with the help of the HEC-RAS program. The normalization carried out on the Deli River in order to flow the discharge is by dredging and raising the dike. Besides the irregular river cross section also made with trapezoidal shape. From the results of Hydraulics Analysis by using HEC-RAS modeling.

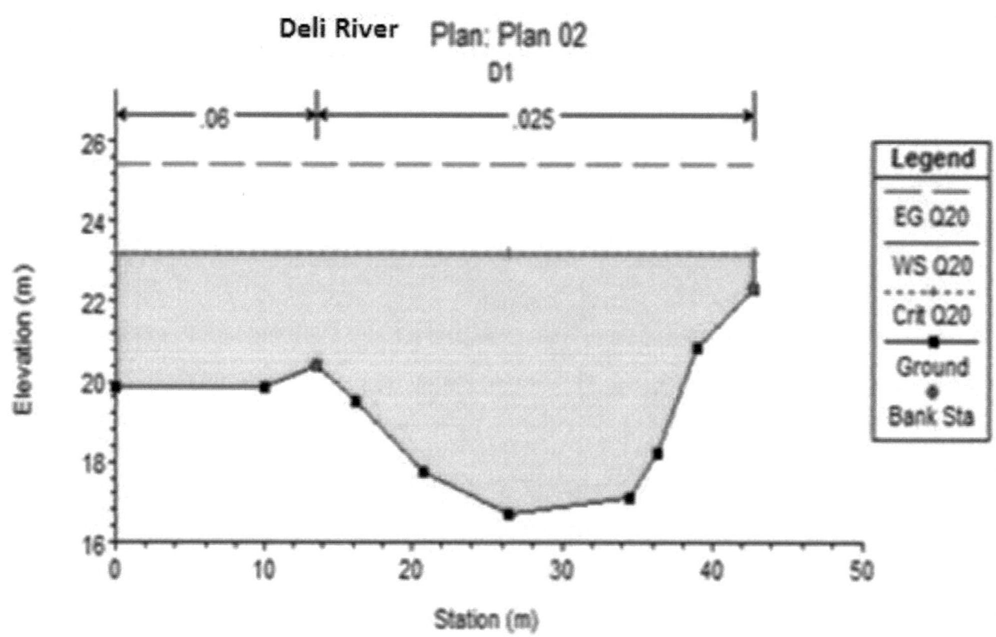

Figure 8: The Runoff Height in the 1,000 Mark in the HEC-RAS Program

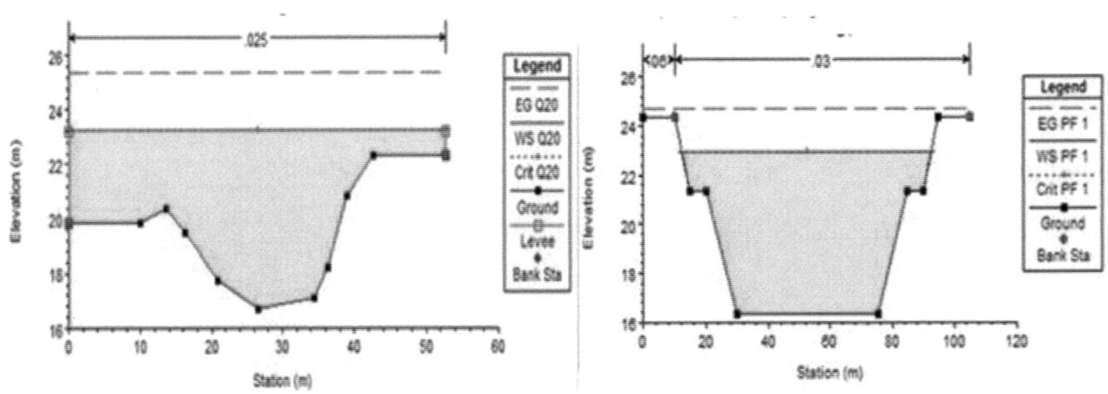

Figure 9: River Condition After Built and Widened Cross Section

3.4 Flood Discharge Index

In the determination of flood index, the calculated discharge is the amount of discharge that causes the flood. So what is meant by the minimum discharge is the smallest discharge that causes the flood starts. Maximum discharge is the largest possible discharge due to the maximum rainfall occurring in the watershed. Based on the maximum and minimum discharge will get the average discharge. Average discharge is the average discharge rate that can cause flooding. The type or the number of variables in the ratio is the same, this is certainly reasonable because it can compare a value of the same object from one or a group of variables. While in its development, the value of the index can be a single value of one variable or the sum of several variables in the same unit about the flood index. The determining index variable is a single variable of the amount of rain in a given value. In this case the value of the flood index still has the value of the unit, but because only one variable that determines, then the determination of the index value is not problematic. The flood index in his dissertation [8] is the ratio between the discharge value and the index value with no units.

In determining the smallest peak discharge that can determine the flood, conducted a rainfall runoff simulation with real time rainfall peak data in each sub watershed. Each simulation of peak rain to one sub watrshed, rain on another sub watershed corresponds to the same real time. The simulation starts from the lowest rain of 1 hour until the rain that can cause flood, so the discharge is called minimum discharge, Q_{min}. The inflow discharge index is formulated as the ratio of the flood discharge that occurs in reducing the minimum discharge with a discharge limit between maximum and minimum. The formula is as follows:

$$IQ = \frac{Qt - Q_1}{Qmaks-} \qquad (7)$$

3.5 Index of Inundation Area

The inundation area is occured in the floodplains due to overflow of water along the modeled river. It takes a great deal of effort to deal with the problem of inundation or floods, such as building floodways according to the needs/capacity planning [9]. The inundation area is a function of the size of the inflow discharge. The bigger the flood discharge, then the inundation that occur will be wider so that the area of inundation caused by the maximum discharge is called the A_{max}, and the extent of the inundation that occurs due to the minimum discharge is called A_{min}. Based on this analysis, the Index of Puddle Area can be formulated as follows:

The classification of the inundation area is adjusted by the flood discharge classification which causes the area of the inundation to obtain two classifications, namely the inundation width between the mean and maximum, and the width of the inundation between the minimum and the average. The inflow discharge index occurs substracted by the minimum discharge with the grace of the discharge between the maximum and the minimum that has been obtained as in Table 4.

Table 4. The Calculation of Inflow Discharge Index (A = 14,8 km^2, L = 7 km)

Return Period	Q_{maks}	Q_{min}	Q_t	Indeks Discharge Inflow (100%)
2	227,08	0,41	10,57	0,04
5	622,55	1,13	18,92	0,03
10	893,66	1,62	24,44	0,03
25	1241,92	2,25	31,43	0,02
50	1503,10	2,72	36,61	0,02
100	1764,09	3,20	41,75	0,02
1000	2635,02	4,78	58,75	0,02

4 Conclusions

Based on the results of direct observation in the field, calculations both technically and program on existing data, the compiler can take some conclusions, namely:

1. Result of rainfall calculation of 5 years return period design with Log Pearson Type III method is 132 mm.
2. Flood Index model on Deli river as flood control in Medan city produce as follows:
   a. After analyzing the 5 years period, the average flood index is 30%. From these results, it can be concluded that the component of Discharge Index is relatively small compared to the composite index of the inundation. The wider the inundation occurs the smaller the Inundation Index (Lt), as well as the smaller the Flood Index.
   b. The direct benefit of this study, especially the case study model of the Deli river, is that the flood index can be estimated easily by knowing the maximum rainfall of the Deli River basin.

Acknowledgements

Acknowledgments are conveyed to DRPM Ministry of Research, Technology and Higher Education through the Research Institute of Universitas Negeri Medan for the funding support. Acknowledgments are also conveyed to Prof. Drs. Motlan, M.Sc.Ph.D. as chairman of the Universitas Negeri Medan Research Institute, who helped carry out the research. Mr. Prof.Dr. Harun Sitompul, M.Pd. as the dean of the Faculty of Engineering, Universitas Negeri Medan, which has given us the opportunity to develop Tri Darma of Higher Education especially in learning.

References

[1] Sudjarwadi. 1997. *Teknik Sumber Daya* Air. Penerbit Universitas Gadjah Mada Yogjakarta.

IOP Conf. Series: Journal of Physics: Conf. Series **970** (2018) 012011 doi:10.1088/1742-6596/970/1/012011

[2] Hasibuan, M. 2005. *Peran Serta Masyarakat dan Kelembagaan Terpadu Dalam Pengelolaan Banjir Di Kota Medan.* Disertasi Doctor, PWD, USU.

[3] Siswoyo, H, 2003,Optimasi Penggunaan Lahan dalam Pengelolaan DAS dengan Pendekatan Aspek Hidrologi berdasarkan teori Hidrograf Satuan, Tesis, ITB

[4] Harahap,R, Mawengkang, 2013. Study on Flood Discharge the Asahan river in North Sumatra Trough HEC-RAS. Prosiding International Conference on BIGSTAR; Bandung 27-28 Oktober 2013. Savoy Homan Hotel Bandung ISBN 978-979-3541-26-6 Polban.

[5] Suripin, 2004, *Sistem Sungai Perkotaan yang berkelanjutan*, Penerbit Andi, Jakarta

[6] Triatmodjo , Bambang (2009), *Hidrologi Terapan*, Beta Offset, Yogyakarta.

[7] Sri Harto,.1993. *Analisis Hidrologi*. Gramedia Pustaka Utama, Jakarta.

[8] Suryadi, Y. 2008. *Metoda Penentuan Indeks Banjir Berdasarkan Fungsi Discharge Puncak Hidrograf Inflow, Luas Genangan, Kedalaman Genangan dan Waktu Genangan*. Disertasi Doktor, Pasca sarjana ITB

[9] Riman, 2011 Evaluasi Sistem Drainase Perkotaan di Kawasan kota Metropolis Surabaya,Widya Teknika Vol.19 No.2; Oktober 2011 , ISSN 1411 – 0660: 39-46 Universitas Widyagama Malang

IOP Conf. Series: Journal of Physics: Conf. Series **970** (2018) 012012 doi:10.1088/1742-6596/970/1/012012

Classification of ECG signal with Support Vector Machine Method for Arrhythmia Detection

Arjon Turnip[*1], M. Ilham Rizqywan[1], Dwi E. Kusumandari[1], Mardi Turnip[2], and Poltak Sihombing[3]

[1]Technical Implementation Unit for Instrumentation Development, Indonesian Institute of Sciences, Indonesia
[2]Technology and Computer Science, Universitas Prima Indonesia, Indonesia
[3]Faculty of Computer Science and Information Technology, Universitas Sumatera Utara, Indonesia

Emails: *jujhin@gmail.com, esti_dek@yahoo.com, marditurnip@unprimdn.ac.id, poltakhombing@yahoo.com

Abstract. An electrocardiogram is a potential bioelectric record that occurs as a result of cardiac activity. QRS Detection with zero crossing calculation is one method that can precisely determine peak R of QRS wave as part of arrhythmia detection. In this paper, two experimental scheme (2 minutes duration with different activities: relaxed and, typing) were conducted. From the two experiments it were obtained: accuracy, sensitivity, and positive predictivity about 100% each for the first experiment and about 79%, 93%, 83% for the second experiment, respectively. Furthermore, the feature set of MIT-BIH arrhythmia using the support vector machine (SVM) method on the WEKA software is evaluated. By combining the available attributes on the WEKA algorithm, the result is constant since all classes of SVM goes to the normal class with average 88.49% accuracy.

1. Introduction

The electrical activity under normal circumstances starts from an impulse formed by a pacemaker at the Sino Atrial (SA) node. The electrical signals from the SA node flow through both atria, causing the two atria to contract to drain blood into the ventricles. Then this electrical signal flows through the Atrio Ventricular (AV) node and then goes to the bunle of His and splits into two the left and right and ends in Purkinjie fibers that activate the ventricular muscle fibers. This causes the two ventricles to contract each other to pump blood throughout the body and produce a pulse. This regular flow of the electricity from the SA node to the AV node causes a regular contraction of the heart muscle known as the sinus beat. Electrocardiograph (ECG) is a device that used to record biological signals formed as a result of the electrical activity of the heart. The ECG is taken by installing electrodes at certain points of the patient's body. The ECG signal has a voltage of up to 3mV and a frequency range of 0.03-100 Hz. The ECG signals have a specific form that can be used as a reference to determine the condition of a person's heart health by a doctor or cardiologist.

The development of medical instrumentation technology based on monitoring systems such as ECG for diagnostic and treatment needs recently tends to switch to wireless technology (ie medical instrumentation that can be used mobile and make patients feel more comfortable) [1-3]. However, some issues related to these demands still need to be developed such as how the tool can be applied

IOP Conf. Series: Journal of Physics: Conf. Series **970** (2018) 012012 doi:10.1088/1742-6596/970/1/012012

easily, quickly, accurately and with the best reliability [4]. One of the wireless-based devices in medical telemetry that it is possible to develop is telecardiology which involves the transmission of electrocardiogram signals (ECG) [5-8]. ECG is a diagnostic tool used to measure and record the electrical activity of the heart in detail. Interpretation of the heart recording signal can be used to analyze the presence of cardiac function that has a vital role in life [8]. With built instruments allowing a patient to be able to measure his heart condition at any time, and the resulting data can be sent to the cloud for subsequent access by interested parties.

Reliable algorithm support in the development of wireless technology with real telecardiology capability is very necessary to provide. The most important thing is to ensure the required quality of the developed ECG system. The conventional wireless applications are designed to optimize average throughput. However, the applied ECG monitoring not only require a moderate data rates, but also require a very high level of reliability. The ECG signals are generated by the stimulation of nerve impulses of the heart. The current on the body surface is used to construct a drop voltage from microvolts into milivolt. The typical of the ECG signals consists of P waves, QRS complexes and T waves. The ECG base voltage is known as the isoelectric line measured as part of tracking following the T wave and predates the next P wave. Our aim is to develop an extraction algorithm that can be used to support the designed of the wireless ECG system.

2. Method

To ensure the patient's environmental feasibility in more humane conditions in both physical and physiological health care, monitoring and recording of their physiologically related data is essential. [10-12]. To reduce the workload of the medical staff and avoid suddenly accidents, an ECG based wireless is proposed. The conceptual arrangement of a wireless ECG medical instrument is shown in Figure 1. The developed system consists of electrode sensors (to measure the ECG signal), a wireless ECG recorded module (to preprocessed and transmit the ECG signals via Bluetooth), and an android mobile device. Then, an ECG monitoring algorithm include the classifier developed in the android mobile device will continuously monitor the subject heart activities. When the abnormal heart rate is detected by classifier, the classified ECG signals will be sent to the cloud or directly to the physician or family. Based on this mechanism, the cardiac state of the subject can be monitored anywhere in the globe as long as the subject under internet network coverage. The main purpose of ECG wireless development is to develop medical equipment with universal and interoperable interfaces such as transparent to end users, easy to use, easy to reconfigure.

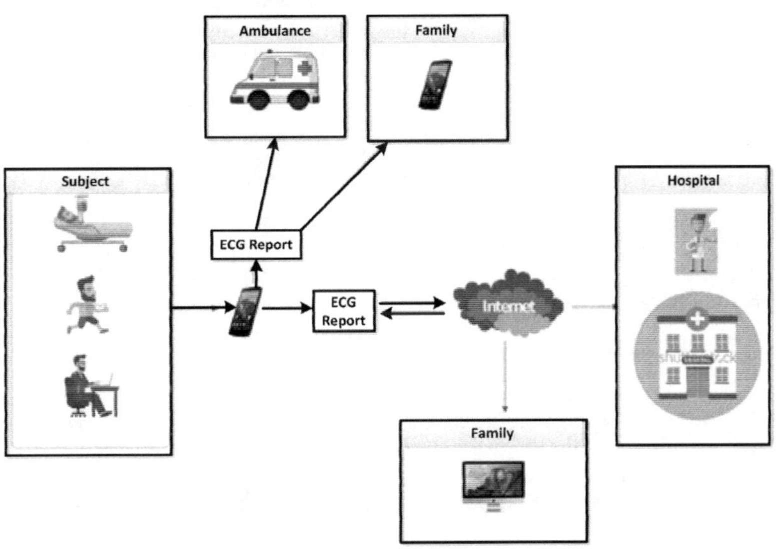

Figure 1. Smart wireless ECG Scheme.

IOP Conf. Series: Journal of Physics: Conf. Series **970** (2018) 012012 doi:10.1088/1742-6596/970/1/012012

In this study, The ECG data is classified using the Support Vector Machine (SVM) method integrated in the Waikato Environment for Knowledge Analysis (WEKA) software. The WEKA is a software that contains Java-based open source data mining applications. The WEKA consists of a collection of machine learning algorithms that can be used to perform generalizations or formulations from a set of sampling data. The SVM WEKA based on norm least mean square (NLMS) extraction are used to forming training vector, establishing SVM WEKA, training the SVM WEKA and diagnosing Arrhythmia. Figure 2 gives the summarized block diagram of the proposed method.

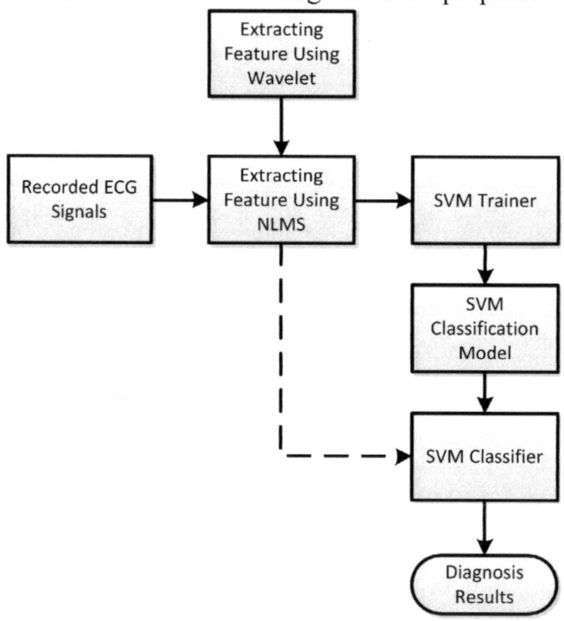

Figure 2. The SVM classifier method based NLMS extraction for ECG Arrhythmia diagnosis.

To evaluate reliability and ensure ease of operation, ECG signals from 24 subjects, aged 24 ± 3 years, were recorded. All subjects are healthy both physically and mentally. Next, each experiment is divided into two session conditions: relax (14 subjects) and type (10 subjects), respectively. Each is equipped with three channel sensors with each location shown in Figure 3 relating to cardiac activity. Once completed, each electrode must be covered with an electrolyte gel to minimize the effect of noise. ECG data were collected from three Ag / AgCl electrodes embedded in a developed wireless ECG system.

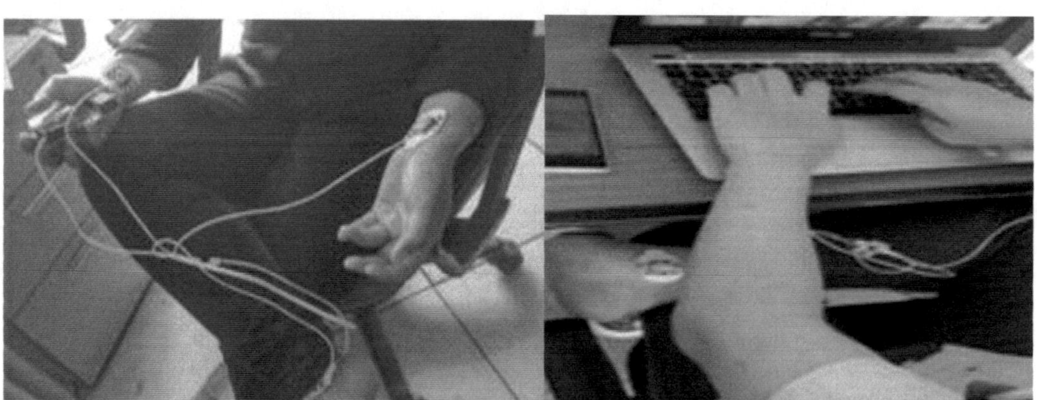

Figure 3. Experiment setup and ECG recorded signals with relax and typing conditions.

3. Signal Processing

During the experiment, the recorded signal is usually mixed with noise. In order to eliminate these signals a filter algorithm with a certain cut-off frequency design is required. The cut-off frequency is an undesired frequency point where there is a decrease in the gain of 3 dB. In designing a filter, the slope steepness level of the filter can be adjusted in such a way that the output as desired. Filters with frequency design is known as bandpass filter. Band pass filter will pass signal with frequencies in certain range and dampen signal with other frequencies. Above and below the resonant frequency resonance frequency, which has an output voltage of 70.7% of the maximum output voltage is the limits that determine the bandwidth of the filter. Signal extraction is an enabling a designed algorithms and its application for extracting information contained in signals. An adaptive filter with Normalized Least Mean Square (NLMS) algorithm is used reduce noise in the recorded ECG signals while extracting the desired signal characteristics [13-16].

Length of Frequency QRS Complex can reach up to 40 Hz, while P and T waves usually have frequency components up to 10 Hz. By using an integrated filter bandpass with adaptive filter (NLMS), the spectral characteristics of the ECG components in the mean and high-frequency parts will be filtered and attenuated. The filtered signal using the proposed method will then be used to determine the temporal-R location with a linear respone. Band Pass Filter with linear respone without integration with adaptive filter then R-wave will usually be more difficult to detect. Complex QRS detection is completed when the R-wave temporal location has been found. If using only one ECG chanel to detect R-wave, the heart's electrical position needs to be considered (the temporal location is determined by the maximum / minimum combination search). It takes a simple decision logic to decide whether to use the maximum or minimum interval position of the R-wave temporal location search.

Feature extraction on QRS detection experiment uses MIT-BIT Arrhytmia training data with WEKA software. In WEKA pre-processing there are attributes that will be omitted one by one or combined. Some of the used extracted features are: Prev indicates the previous data(Rn-Rn-1), Next indicated the next processed data (Rn + 1 - Rn), Local_Avg: Average before and after designated data as much as W (Width), Avg: Average of all Next in 1 data record, RRIR: Comparison of Next result designated divided Next in front (Nextn / Nextn + 1), 10 RRIR: Comparison of Next to 10 designated results on the next data.

The existing SVM classifier embended on the WEKA algorithm is used for classification. The SVM is a learning system that uses hypothetical space in the form of linear functions in a high-dimensional feature space, trained with learning algorithms based on optimization theory by applying inductive bias (derived from the theory of statistical learning). With a simple concept, it can be explained as the search for the best hyperplane that serves as a separator of two classes in the input space (i.e., a member of two classes: +1 and -1 and shares an alternate line of discrimination boundaries). Margin is the distance between the hyperplane and the nearest pattern of each class. The closest pattern is called a support vector. The effort to locate the hyperplane is at the heart of the learning process in the SVM.

4. Results and Discussions

The algorithm has been tested on the recorded data from ten healthy volunteers subject aged around 25 ± 3 years old and it functions normally. Figures 4 and 5 show the recorded using developved system (top) and filtered signals using adaptive filter with normalized mean square algorithm (bottom). In normal ECG, the time interval between R-R is 0.6-1s, in case of fast heartbeat the time interval is less 0.6s which is known as sinus tachycardia; in case of slow heartbeat the time interval is more than 1 sec that is known as sinus bradycardia. By referring Figures 4 and 5 we can say that the R-R interval for normal case is between 0.6-1s. All signals from all subject and different distance of wireless ECG recording signals are in the standard form of ECG signals. Compared with the data measured by medical instruments in nursing centers and hospitals, our physiological status monitoring system has

IOP Conf. Series: Journal of Physics: Conf. Series **970** (2018) 012012 doi:10.1088/1742-6596/970/1/012012

reached the application level. By improving the signal processing ability for filtering, the quality of filtered signal from a farther distance.

Initial stage on detecting subject's heart condition is developing an algorithm that can be used to detect the QRS position as in Figure 6. The QRS positioning information on each recording signal will make it easier for doctors or family to know their heart conditions. To enhance the accuracy of peak detection related to the QRS position, the signal is preprocessed using an adaptive filter. The R and S points including RR interval were well detected as shown in Figure 7. The obtained results clearly different from the recorded data of relax than typing subjects. If in the previous results with a general filter algorithm the difference is highly significant but with the adaptive filter the peak detection accuracy is only slightly different. Then the proposed filter is sufficiently able to overcome the dynamic interference on the signal recording.

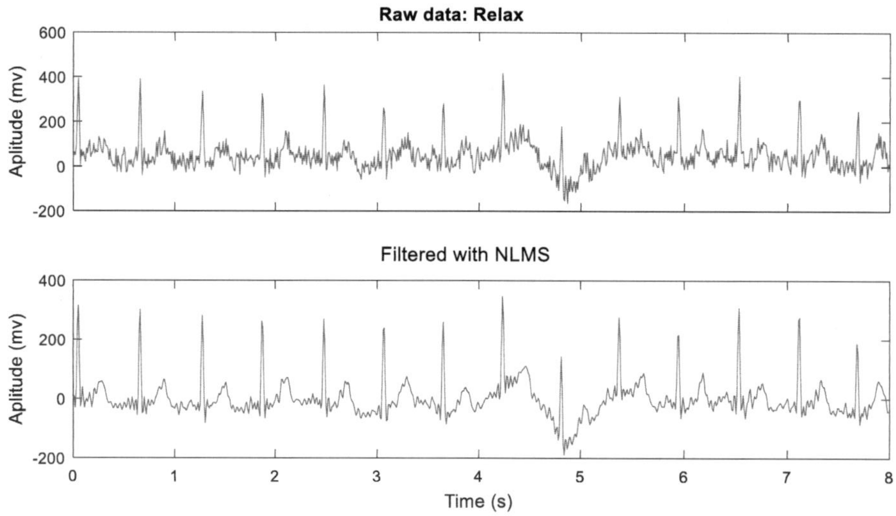

Figure 4. Filtered ECG signals from relax subject.

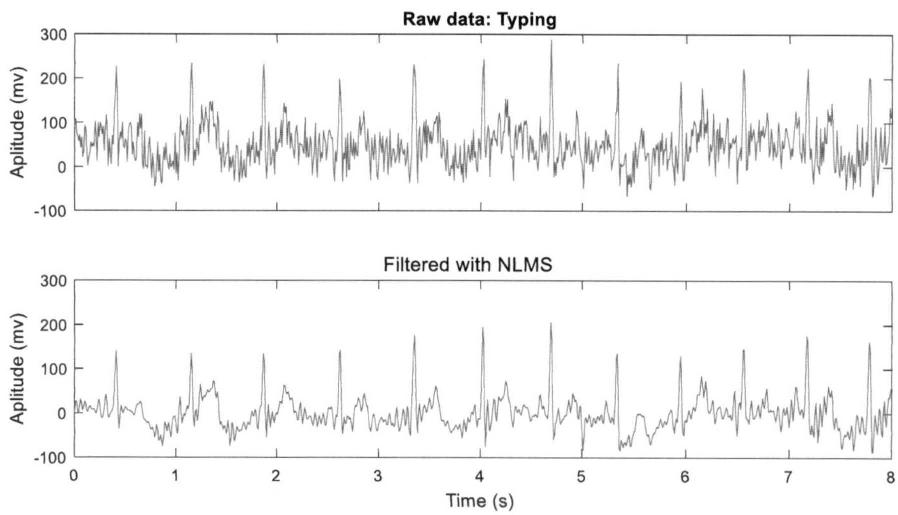

Figure 5. Filtered ECG signals from typing subject.

The QRS Detection is the name for a combination of three visible graphical deflections on the ECG signals (usually represented in the center and most visually clear of the search). This corresponds to the depolarization of the right and left ventricular human heart. In adults, it usually lasts 0.06-0.10 seconds; In children and during physical activity, may be shorter. For the results in the first experiment the QRS Detection was relaxed against 14 subject all peaks were detected well and had 100% accuracy and sensitivity as shown in Table 1. This perfect result is obtained because the subject at the rilek condition, ECG signal recording process tends not influenced by interference. The results of the experiment on 10 subjects during typing, QRS detection is very diverse. This is caused by the difference in frequency when the subject is typing so that not all results achieve 100% sensitivity and accuracy as shown in Table 2. The SVM uses the existing training data on Arryhtmia's MIT-BIH by using feature sets to be removed and combined so as to obtain stable results for SVM classification in WEKA software, there are Correct and Incorrect indicators on SVM calcification results which show the accuracy of classification. The SVM classification results show that all values pointing to other value and Classifier indicator has a value of 0 which indicate that the accuracy of the feature set having a constant accuracy.

Table 1. The QRS detection results for relax condition.

No.	File	Desired peak	Detected Peak	Undetected Peak	Undesired peak	TN	TP	FN	FP	Accuracy	Sensitivity	PP
1	file1	169	169	-	0	0	169	0	0	1	1	1
2	file2	167	167	-	0	0	167	0	0	1	1	1
3	file3	156	156	-	0	0	156	0	0	1	1	1
4	file4	165	163	-	0	0	163	0	0	1	1	1
5	file5	150	150	-	0	0	150	0	0	1	1	1
6	file6	190	190	-	0	0	190	0	0	1	1	1
7	file7	190	190	-	0	0	190	0	0	1	1	1
8	file8	177	177	-	0	0	177	0	0	1	1	1
9	file9	166	166	-	0	0	166	0	0	1	1	1
10	file10	223	223	-	0	0	223	0	0	1	1	1
11	file11	169	169	-	0	0	169	0	0	1	1	1
12	file12	176	176	-	0	0	176	0	0	1	1	1
13	file13	220	220	-	0	0	220	0	0	1	1	1
14	file14	152	152	-	0	0	152	0	0	1	1	1
Total		2470	2468	0	0	0	2468	0	0	1	1	1

Table 2. The QRS detection results for typing condition.

No.	File	Desired Peak	Detected Peak	Undetected Peak	Undesired Peak	TN	TP	FN	FP	Accuracy	Sensitivity	PP
1	file1	201	177	4	20	0	177	4	20	0,8806	0,977901	0,89848
2	file2	235	161	19	55	0	161	19	55	0,68511	0,894444	0,74537
3	file3	154	146	1	7	0	146	1	7	0,94805	0,993197	0,95425
4	file4	158	134	7	17	0	134	7	17	0,8481	0,950355	0,88742
5	file5	160	122	15	23	0	122	15	23	0,7625	0,890511	0,84138
6	file6	251	139	24	88	0	139	24	88	0,55378	0,852761	0,61233
7	file7	192	159	4	29	0	159	4	29	0,82813	0,97546	0,84574
8	file8	211	165	11	35	0	165	11	35	0,78199	0,9375	0,825
9	file9	161	147	2	12	0	147	2	12	0,91304	0,986577	0,92453
10	file10	186	144	9	33	0	144	9	33	0,77419	0,941176	0,81356
Total		1909	1494	96	319	0	1494	96	319	0,79755	0,939988	0,83481

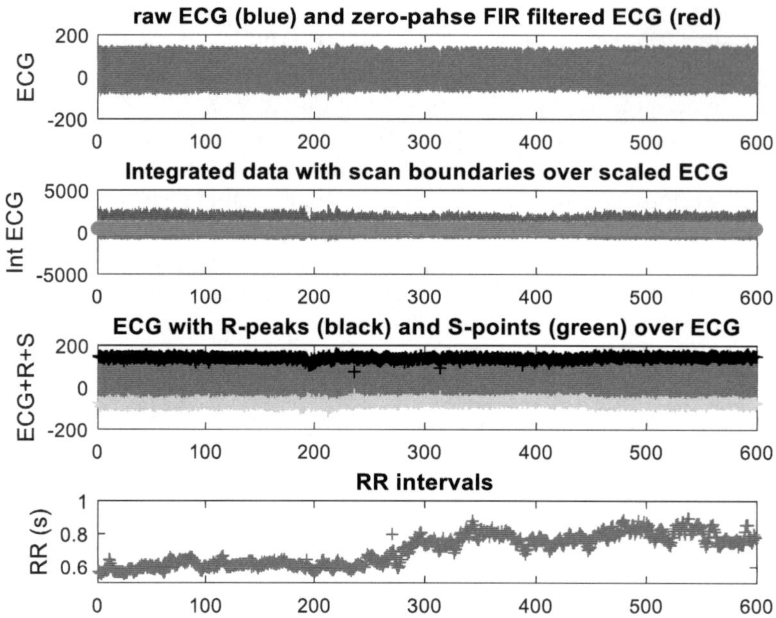

Figure 6. Detected R & S point and RR Interval.

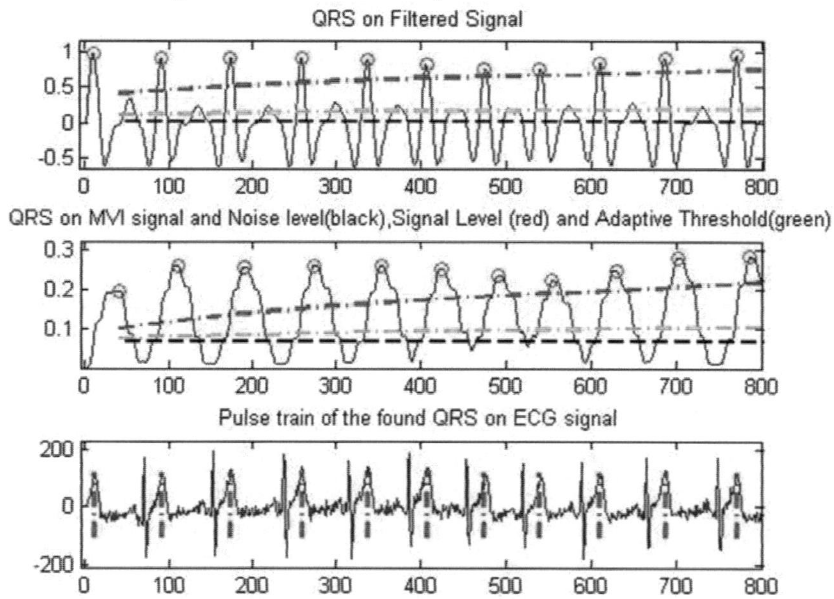

Figure 7. Detected QRS for relax condition

5. Conclusions

In this paper, the developed (i.e., wireless ECG system that consists of mobile physiological examination device and wireless base station) and its application in the experiment is evaluated. One of the crucial steps in the ECG analysis is to accurately detect the different form of P, Q, R and S which represent the entire heart cycle. An application of the adaptive filter with the developed system, the higher qualitiy of the ECG signals is achieved. An improvement of QRS peak detection accuracy supported by adaptive filter about 86% is achieved.

Acknowledgment
This research was supported by the Technical Implementation Unit for Instrumentation Development, Indonesian Institute of Sciences and, Insentif Riset Sistem Inovasi Nasional (INSINAS), Kementerian Riset, Teknologi, dan Pendidikan Tinggi (Ristekdikti), Indonesia.

References

[1] U. Varshney, Pervasive Healthcare Computing—EMR/EHR,Wireless and Health Monitoring. New York: Springer, 2009.

[2] S. Scalvini, G. Martinelli, D. Baratti, D. Domenighini, M. Benigno, L. Paletta, E. Zanelli, and A. Giordano, "Telecardiology: One-lead electrocardiogram monitoring and nurse triage in chronic heart failure," J. Telemedicine Telecare, vol. 11, Suppl. 1, pp. 18–20, Jul. 2005.

[3] A. Alesanco and J. Garcıa, "Clinical assessment of wireless ecg transmission in real-time cardiac telemonitoring," IEEE Trans. Inf. Technol. Biomed., vol. 14, no. 5, pp. 1144–1152, Sep. 2010.

[4] S. D. Baker, "Building reliability into your digital telemetry system," Health Manag. Technol., vol. 23, no. 6, pp. 100–103, Jun. 2002.

[5] S. Sneha and U. Varshney, "A wireless ECG monitoring system for pervasive healthcare," Int. J. Electron. Healthcare, vol. 3, no. 1, pp. 32–50, 2007.

[6] A. Turnip, M. I. Rizgyawan, and D. E. Kusumandari, "Application of Support Vector Machine Classifier on Developed Wireless ECG System," International Journal of Pharma Medicine and Biological Sciences Vol. 5, No. 3, 2016.

[7] D. Cypher, N. Chevrollier, N. Montavont, and N. Golmie, "Prevailing over wires in healthcare environments: Benefits and challenges," IEEE Commun. Mag., vol. 44, no. 4, pp. 56–63, Apr. 2006.

[8] K. Kang, K.-J. Park, J.-J. Song, C.-H. Yoon, and L. Sha, "A medicalgrade wireless architecture for remote electrocardiography," IEEE Trans. Inf. Technol. Biomed., vol. 15, no. 2, pp. 260–267, Mar. 2011.

[9] D. Olguin, Frantz Bouchereau and Sergio Martinez; "Adaptive notch filter for ECG signals based on the LMS algorithm with variable step-size parameter" conference on information sciences and systems, the John Hopkins University; March 16 –18, 2005.

[10] P. Johnson and D. C. Andrews, "Remote continuous physiological monitoring in the home," J. Telmed. Telecare, vol. 2, no. 2, pp. 107–113, 1996.

[11] M. J. Rodriguez, M. T. Arredondo, F. del Pozo, E. J. Gomez, A. Martinez, and A. Dopico, "A home telecare management system," J. Telmed. Telecare, vol. 1, no. 2, pp. 86–94, 1995.

[12] P. Varady, Z. Benyo, and B. Benyo, "An open architecture patient monitoring system using standard technologies," IEEE Trans. Inf. Technol. Biomed., vol. 6, no. 1, pp. 95–98, Mar. 2002.

[13] A. Gacek and W. Pedrycz, ECG Signal Processing, Classification and Interpretation, Springer, 2012.

[14] Z. Jiashu, T. Heng-Ming, "Adaptive Noise Cancellation Algorithm for Speech Processing", IEEE Transactions, pp 2489-2492, 2007.

[15] H. Jigram, Shah, J. M. Joshi, "Digital signal processing" University science press laxmi publication company Limited New Delhi

[16] A. Turnip, and I. R. Setiawan, "Artifacts Removal of EEG Signals using Nonlinear Adaptive Autoregressive," International Journal of Information and Electronics Engineering, vol. 5, no. 3, May 2015.

[17] A. Turnip and M. Siahaan, Adaptive Principal Component Analysis based Recursive Least Squares for Artifact Removal of EEG Signals, Advanced Science Letters, vol. 20, no.10-12, pp. 2034-2037(4), October 2014.

[18] A. Turnip and K.-S. Hong, "Classifying mental activities from EEG-P300 signals using adaptive neural network", International Journal of Innovative Computing, Information and Control (IJICIC), vol. 8, no. 7, pp. 6429-6443, 2012.

IOP Conf. Series: Journal of Physics: Conf. Series **970** (2018) 012013 doi:10.1088/1742-6596/970/1/012013

Design of Simulation Product for Stability of Electric Power System Using Power System Stabilizer and Optimal Control

Agus Junaidi[1], Abdul Hamid K[2]
[1]Department of Electrical Engineering Faculty of Engineering Universitas Negeri Medan
[1]Graduate student- Doctoral Program Faculty of Engineering Universitas Negeri Padang
[2]Faculty of Engineering Universitas Negeri Medan

mhabib4n01@yahoo.com

Abstract. This paper will discuss the use of optimal control and Power System Stabilizer (PSS) in improving the oscillation of electric power system. Oscillations in the electric power system can occur due to the sudden release of the load (Switcing-Off). The oscillation of an unstable system for a long time causes the equipment to work in an interruption. To overcome this problem, a control device is required that can work effectively in repairing the oscillation.The power system is modeled from the Single Machine Infinite Bus Model (SMIB). The state space equation is used to mathematically model SMIB. SMIB system which is a plant will be formed togetherness state variables (State-Space), using riccati equation then determined the optimal gain as controller plant. Plant is also controlled by Power Stabilizer System using phase compensation method. Using Matlab Software based simulation will be observed response of rotor speed change and rotor angle change for each of the two controlling methods. Simulation results using the Simulink- MATLAB 6.1 software will compare the analysis of the plant state in Open loop state and use the controller. The simulation response shows that the optimal control and PSS can improve the stability of the power system in terms of acceleration to achieve settling-time and Over Shoot improvement. From the results of both methods are able to improve system performance.
Index-Terms: Power System Stabilizer (PSS), optimal control, Simulation.

1. Intoruction

The power system generally consists of several generating units, transmission lines and various load centers. In operation, the power system can not be separated from interference, be it interference that is temporary or permanent disturbance. If one generator or another generator is disturbed, for example caused by a load change or discharged the generator may cause changes to certain variables on the system, such as voltage, frequency, and so on. Any such changes will affect the stability of the electric power system. The stability of a power system is defined as the ability of a power system to maintain synchronization and balance in the system due to interference [1,7].

A proper and continuous stability study is necessary to analyze the system in order to work effectively. To study the dynamic stability used modeling of components such as genarator, transmission line and load. Modeling is derived from mathematical equations in the form of linear differential equations to represent the dynamics behavior of the system. The application of the riccati matrix equation to determine the optimum gain fed to the excitation system is an alternative method of electrical system stability. This problem is very interesting to observe because with the application of riccati matrix equation is expected to give damping oscillation of electric power system is good.

Power Stabilizer System (PSS) in this paper is also discussed with the phase compensation method installed in the exit system.

Stability problems in the power system are related to synchronous machine behavior after interruption. Instability of the electric power system will result in a shock on the rotor shaft of the generator caused by the oscillation of the engine itself or the oscillation due to interaction between machines. If this happens continuously will result in generator off, or it may even lead to the breaking of the generator rotor. Basically, the stability is divided into steady state stability and transient stability [2]. Transient stability is associated with sudden large disturbances, such as short circuit interruptions, channel shutdown, removal or disconnection of loads. While steady state stability is related to the ability of the power system to return to operating point condition after a minor disturbance such as power or load changes. Steady state stability is also called dynamic stability. The stability study, to determine whether the electrical system is still stable or not after the interference, it is observed through the system state variable as an indicator of stability, including the angle of the rotor swing (δ), and the rotation speed of the rotor (ω).

2. Single Machine Infinite Bus Model

The electrical power system model used in the paper is based on the analysis of SMIB (Single Infinite Bus Machine), a simultaneous engine connected to an infinite bus through a transmission line, as shown in Fig. 1, with *G* the engine simultaneously, and respectively as a resistance *(R)* and an equivalent reactance *(X)*, and as an infinite bus voltage *(Vb)*

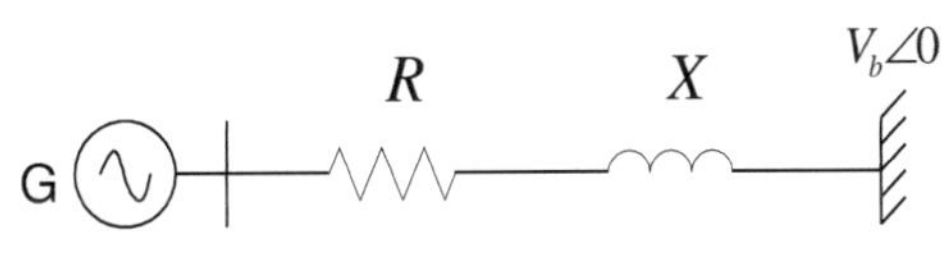

Figure 1. Single Machine Infinite Bus (SMIB) Model

The state variable state of the electric power system is given by [3]:

$$\begin{bmatrix} \Delta\dot{\delta} \\ \Delta\dot{\omega} \\ \Delta\dot{E}_q' \\ \Delta\dot{E}_{fd} \end{bmatrix} = \begin{bmatrix} 0 & \omega_b & 0 & 0 \\ -\frac{K_1}{M} & -\frac{D}{M} & -\frac{K_2}{M} & 0 \\ -\frac{K_4}{T_{d0}'} & 0 & -\frac{1}{T_{d0}'K_3} & \frac{1}{T_{d0}'} \\ -\frac{K_A K_5}{T_A} & 0 & -\frac{K_A K_6}{T_A} & -\frac{1}{T_A} \end{bmatrix} \begin{bmatrix} \Delta\delta \\ \Delta\omega \\ \Delta E_q' \\ \Delta E_{fd} \end{bmatrix} + \begin{bmatrix} 0 \\ 0 \\ 0 \\ \frac{K_A}{T_A} \end{bmatrix} \begin{bmatrix} u_E \end{bmatrix} \quad (1)$$

The power system diagram block shown in Figure 2, consists of two main loops, mechanical loops and electrical loops. The mechanical loop has two function block switches from left to right. The first block is based on the equation of torque balance, and the second block shows the angular and velocity relationship for the selected unit. In this block (M) is an inertial constant, mechanical damper coefficients *(D)*, and simultaneous velocity ($2\pi f$). The electric loop has two blocks of function transfer from right to left. The first block represents the system of regulation of voltage and excitation with time constants (T_A) and a total Gain (K_A). The second block represents the field override function as the effect of armature reaction, with constant effective time ($T_{d0}'K_3$) and gain (K3).

IOP Conf. Series: Journal of Physics: Conf. Series **970** (2018) 012013 doi:10.1088/1742-6596/970/1/012013

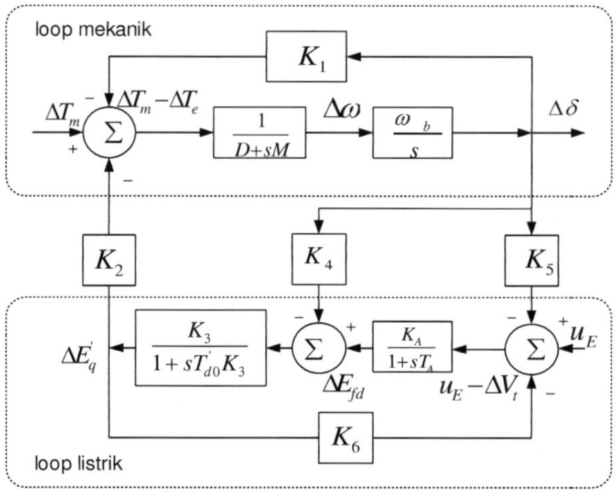

Figure 2. Power system diagram bloc

3. Controller Design
Matrix riccati Equation

The system expressed by the state space equation:

$$\dot{x} = Ax + Bu \tag{2}$$

can be formed into riccati matrix as follows [4];

$$A^T P + PA + Q - PBR^{-1}B^T P = 0 \tag{3}$$

$$K = R^{-1}B^T P \tag{4}$$

Phase Compensation Method

Power System Stabilizer (PSS) is generally installed on the excitation system [5] of the speed response with the form of mathematical equations [6] ie:

$$G_s(s) = \left[\frac{K_o \tau_o s}{1+\tau_o s}\right]\left[\frac{1+a\tau\ s}{1+\tau s}\right]^n \tag{5}$$

Previously we need to calculate the characteristic equation, undamped natural frequency and damping factor of inertia system.

$$d(s) = s^2 + 2\zeta_x \omega_x + \omega_x^2 \tag{6}$$

With:

$$\omega_x = \sqrt{K_1 \omega_R / 2H}$$
$$\zeta_x = D / 4H\omega_n \tag{7}$$

4. Methodology

a. Research data.

The power system data as a controlled plant is taken from reference [5].

b. Research Tools.

Research tool in the form of hardware (Hard Ware) PC intel Cel. 2.00 Gg Ram 384 Monitor 17 "and use software (Soft Ware) Matlab 6.1 / Simulink

IOP Conf. Series: Journal of Physics: Conf. Series **970** (2018) 012013 doi:10.1088/1742-6596/970/1/012013

C. Stages Of Research

Stages of the research implementation is given in the flow diagram below.

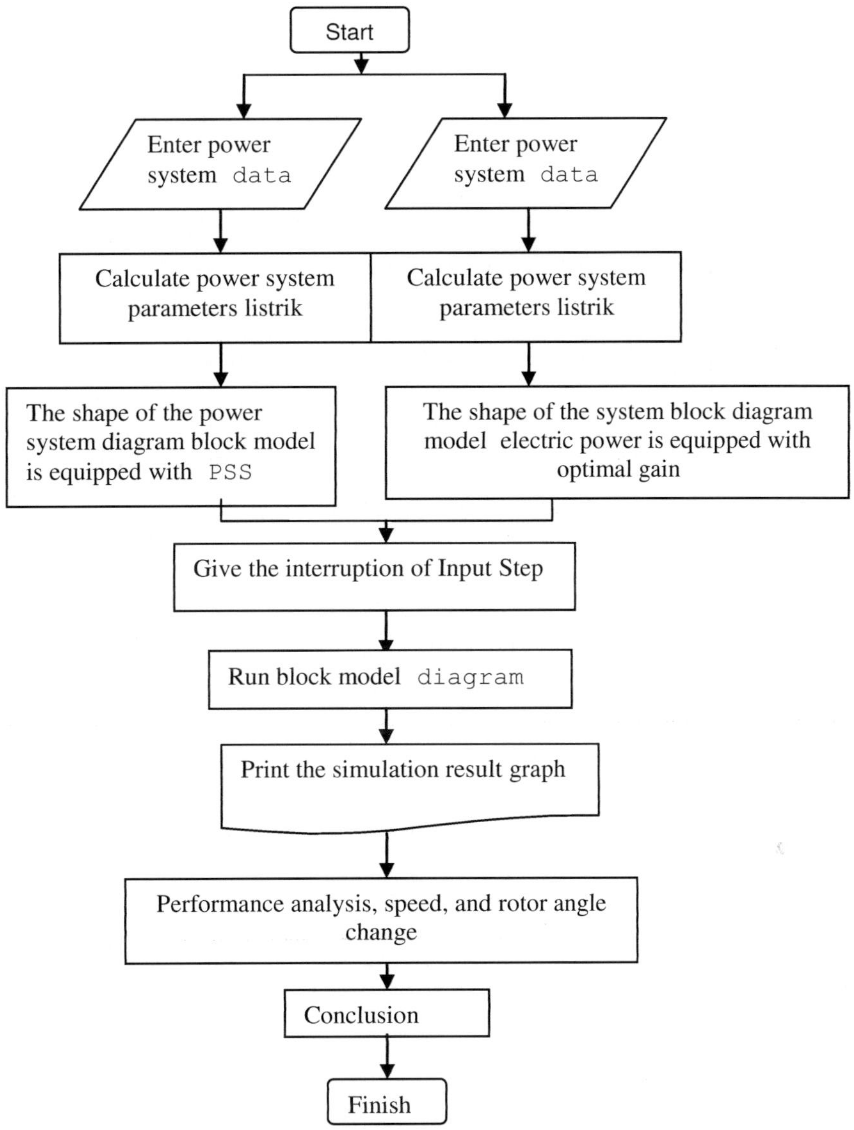

Figure 3. Stages of the research

Stages of research as shown in Figure 3. Production of simulation run by entering data of power system at model of SMIB. SMIB is treated with optimal control and PSS. The simulation of SMIB is run using simulink matlab. Simulation result of SMIB observed oscillation of system stability, seen overshoot and overspeed through display of simulation result

IOP Conf. Series: Journal of Physics: Conf. Series **970** (2018) 012013 doi:10.1088/1742-6596/970/1/012013

5. Results and Discussion

The stability simulation of the electric power system uses optimal control and PSS is shown in Figure 4

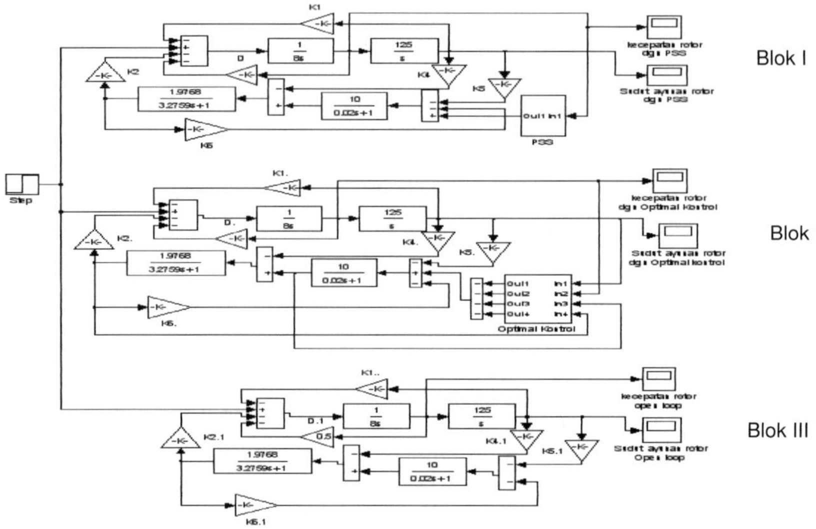

Figure 4. Simulation circuit

Figure 4 is a single machine simulation circuit connected to SMIB system using a simulink. Block 1 is a SMIB diagram controlled by PSS while the block2 SMIB diagram is controlled by Optimal Gain, while Block 3 is an open loop SMIB system. Create sub system of PSS and Optimal Controls are shown in Figures 5a and 5b.

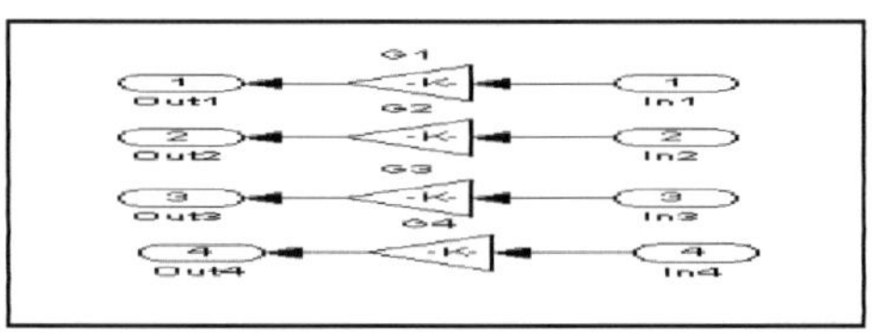

Figure 5a. Sub-system of optimal control

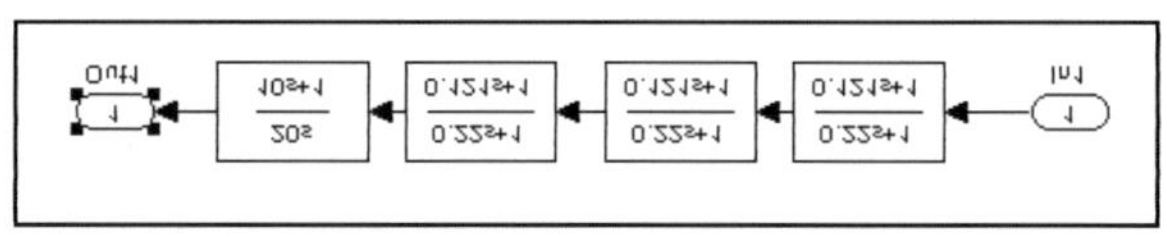

Figure 5b. Sub-system of PSS

Plotting Simulation Result

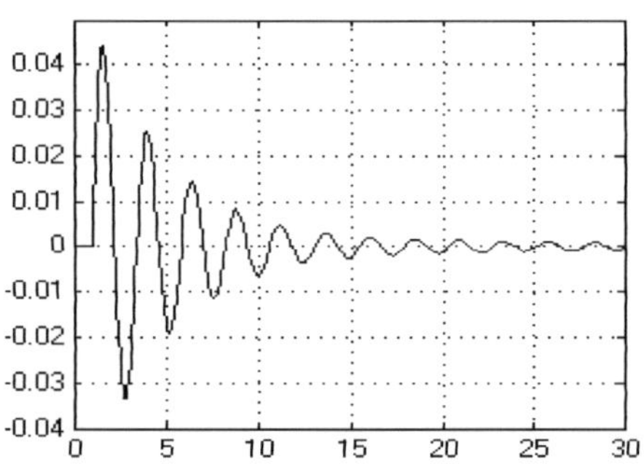

Figure 6a. Open Loop Response Of ω

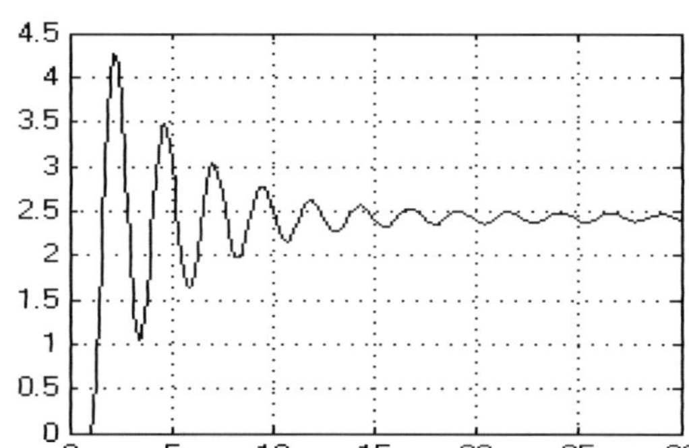

Figure 6.b. Open loop response of δ

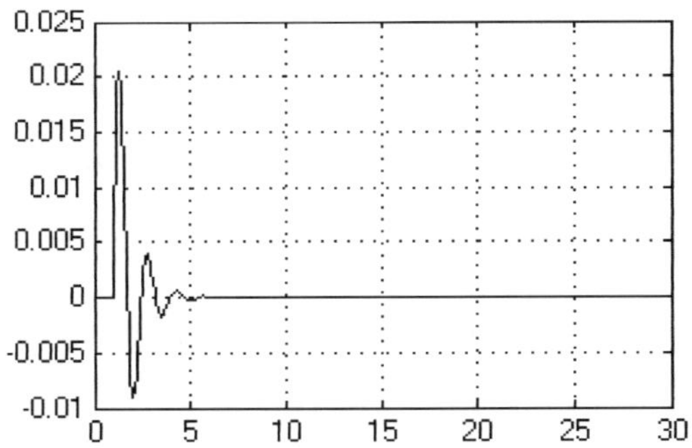

Figure 7a. Close Loop Response ω Using *PSS*

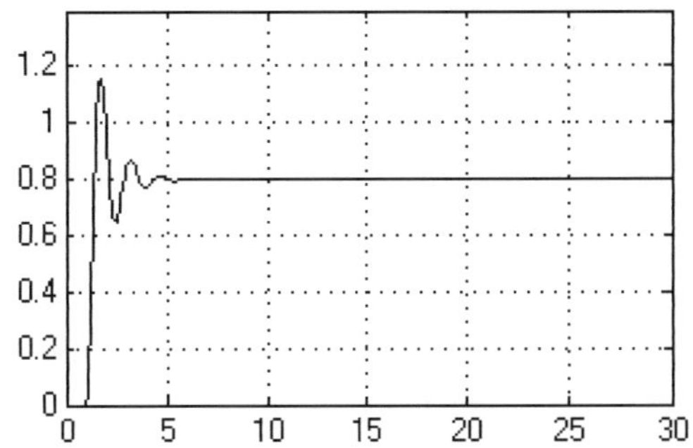

Figure 7b. Close Loop Response δ Using *PSS*

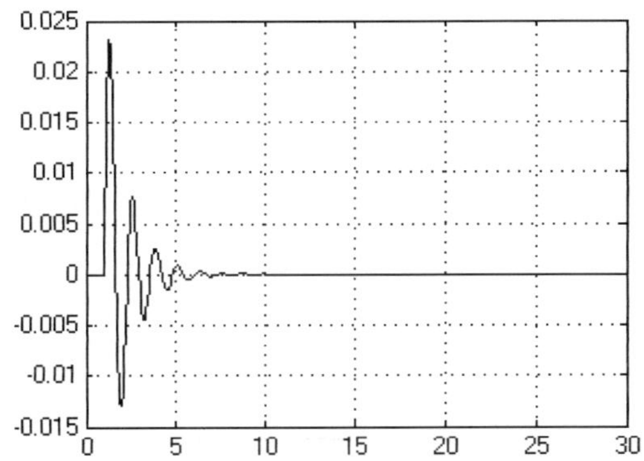

Figure 8a. Close Loop Response ω Using Optimal Control

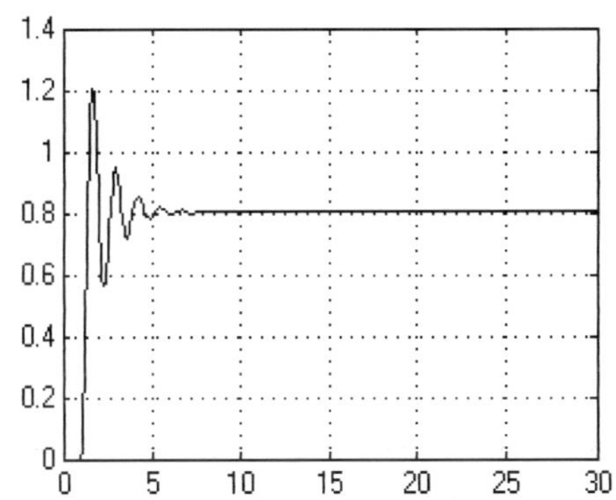

Figure 8b. Close Loop Response δ Using Optimal control

Application of Optimal control using Linear Quadratur Regulator (LQR) and PSS with Matlab 6.1 Simulation shows that the resulting response is able to mered oscillation and improve stability compared to open loop condition. Performance improvements can be seen from the reduction of Over Shoot and settling time on the test in the form of input step as a mechanical input. Power System Stabilizer (PSS) reduces over shot and settling time to change the rotor speed from 0.042 to 0.02 and reach steady state condition within 5 seconds, while for rotor speed change from 4.3 to 1.2 and reach steady state condition at 6 seconds. Improvements made by Optimal Control dampen over shoot for rotor angle change from 4.3 to 1.3 and reach steady state at 7.5 seconds while the rotor speed changes from 0.042 to 0.22 and reach steady state at 7.5 seconds.

6. Conclusion.

Model of electric power system formed linear differential equations then formed in the equation State variable by applying Power System Stabilizer and Optimal Control can be used for dynamic stability analysis. Implementation of installation Optimal control and PSS able to reduce over shoot and settling time so that system performance becomes better than open loop system. Both controllers are able to properly repair the oscillations in the electric power system.

Reference

[1] Stevenson, W.D., dan Idris,K., *Analisis SistemTenaga Listrik*, Erlangga, Jakarta, 1990.

[2] Kundur P. "Power System Stability and Control", McGraw-Hill, 1994 *Control,* IEEE service Center Single Publication Salea dept. 1980

[3] Padiyar, K.R., *Power System Dynamis Stability and Control*, pp. 191 – 268, John Wiley & Sons , Singapore, 1996.

[4] Katsuhiko Ogata, *Designing Liniar Control Systems With matlab"* Prentice hall, ltd. © 1994

[5]. KE. Bollinger,"*IEEE Tutorial Course Power Systems Stabilization Via Exitation*

[6] P.M. Anderson , A.A Fouad *" Power System Control and stability"* The Iowa State university Press, © 1977

[7] Adi Soeprijanto, *" Metode Sederhana Penalaan power Sytems Stabilizer"* Tesis Magister Teknik pengaturan ITB, 1994

IOP Conf. Series: Journal of Physics: Conf. Series **970** (2018) 012014 doi:10.1088/1742-6596/970/1/012014

The effect of Iron Supplying on VO2 MAX and Haematology Parameter on Menstrual Woman

Rika Nailuvar Sinaga[1], Novita Sari Harahap[1a], Rima Mediyana Sari [1]

[1] Faculty of sport science, Medan State University (Unimed), Indonesia

[a]Corresponding author's: novita.s2102@gmail.com

Abstract. One of the supporting factors to have good aerobic endurance requires high VO2 max levels supported by good quality and quantity level of Haematology parameter especially such as erythrocytes, hematocrit and hemoglobin. One of the components in hemoglobin is iron which functions as theoxygen transport to parts of all body required in the process of metabolism. The objective of this research was to find out the difference between VO2 Max and Haematology parameter between iron supplying and no iron supplyingonmenstrual woman. The type of this research is quasi experimental research with non-randomized control group Pretest-Postest Design.The subjectsarethe studentsat faculty of Sports Sciences, Medan State University with the criteria of female gender, monthly regular menstrual cycle, having the level of health and the level of training, willing to be a sample by filling out informed consent. The total number of research subjectsis twenty students, divided into two groups namely the treatment group and the control group. The hematology parameter was measuredby Haemotology analyzer and VO2 Max was measured by multi-stage run. The result showed that there was a significant effect of iron supplyingon the increase of erythrocyte level on menstrual women, hemotocrit, haemoglobin and an increase in VO2 Max. Iron supplying on menstrual woman has the effect on the increase of erythrocyte, hematocrit, hemoglobin level and VO2 Max

1. Introduction

Nowadays women's participation in match / competition especially in sport has been increasing. Biologically, women will have menstrual periods every month where this cycle is continuously and regularly will be passed. Hinton et al (2000) says that a woman in menstrual period will lose bloodwhich affectsthe athlete performance.

Healthy physic is one of the supporting factors to achieve maximum performance in sport because it has good cardiovascular resistance (aerobic). To have a good aerobic endurance requires a high level of VO2 max. Many factors affect VO2 max such as heart, lung, Haematology, blood vessels, andskeletalmuscle in consuming oxygen. There are many components contained in Haematology such as Erythrocytes, one of the blood cellscontained Hemoglobin. Hematocrit is the ratio of red blood cells to blood volume. If one of these components has a low capacity, it will affect the level of VO2max (Fox, 1988) especially Hemoglobin. (Zhu and Haas, 1997) also said that a decrease in VO2 max can occur in anemic patients with decreased Hemoglobin levels and consequently the oxygen transport capacity in the blood will decreases.

Haas and Brownlie (2001) said thatiron is a mineral in hemoglobin namely a protein found in red blood cells (erythrocytes). Iron functions as the formation of red blood cells and these minerals provide a lot of functions in the oxygen transport to parts of all body required in the body metabolic processes. Therefore, iron is very necessary for female athletes during menstruation. According to Weaver and Rajaram (1992), Iron is widely used as one of the additional minerals for female athletes in performing daily physical exercise. In addition, female athletes in sports(endurance) also require more iron supplementation when compared with women who are less active to obtain more energy in doing heavy work (Fogelholm, 1995).

Dueck, Manore and Matt (1996), says that a female athlete in menstrual period can't perform better than before menstrual period. This problem can affect the level of cardiovascularendurance to maximum work intensity.

Newhouse and Clement (1988) says that the current problem of iron deficiency inbody is caused by bleeding of menstruation taking place in many female athletes. Inadequate iron consumption causes reducing oxygen conveyed to the muscle tissues. This problem arises when female athletes do not consume adequate iron in their main course at the time of menstruation, therefore blood carrying oxygen decreases and can affect athlete performance.

Roseann et al (1992), says that in addition to the insufficient amount of iron consumed, women have a necessity for increased iron. The loss of blood due to menstrual periods from month to month is commonly a daily iron loss about 1-2 mg or monthly iron loss about 28 mg. women who do not have menstrual period also lose 0.5-1mg iron in a day most women do not consume the amounts of ironsufficiently (Clement et al,1982). There are only 6 mgirons in every 1000 calories of well-chosen food on a daily main course. A woman who is menstruating must eat 3000 calories to meet daily iron need until18 mg. The Problem arises when many women only take from 1500 to 1600 calories each day. Therefore, the iron formed to be food(supplement) bit help to fulfill the inadequate iron need in body. Considering background above it is necessary to examine the effects of iron supplement onHaematologyparameter and VO2 max on menstrual woman.

2. Subject and Methods

2.1. *Sample*

The sample in this research is 20 students at Faculty of Sport Sciences, Medan State University. The sample has several criteria that must be fulfilled as following: female sex, monthly regular menstrual cycle, having level of health and level of training, willing to be sample by filling out informed consent.

2.2. *Research design*

This research is a quasi experimental research with design of Non Randomized Control Group Pretest-Postest Design. The sample was divided into two groups, each group consists of 10 people. The first group (P1) as a control group was given a placebo in the form of starch and second group (P2) as the treatment group was given iron. The iron and starch was given once a day, for 7 days before the date of menstruation. The Parameter examination of Haemotology and VO2 max was conducted on the third day of menstruation.

2.3. *Measurement of VO2 max*

Measurement of VO2 max used a multi-stage run test or bleep test. The bleep test procedure is to measure a distance of 20 meters and mark it on both tips with a cone or other mark as distance, thenprepare the cassette and the tape recorder that serves a rhythm when running, have the participants warm up and stretch, turn on the tape and then get all participants to perform a multi-stage run (bleep test) until the participants can no longer follow the rhythm that has been specified. After that take a note level of the sample capability and be measured VO2max (compared with table).

2.4. *Complete Blood Measurement*

The blood sampling for Haemotology parameters examination wasperformed on the third day of menstruation, then measurements were made using Haematology Analyzer.

2.5. Data analysis
Data obtained from the research were analyzed statistically using unpaired t-test with significance level below 0.05 ($p < 0.05$) and 95% confidence level ($\alpha = 0.05$). The first step is the normality test and homogeneity test. Data were analyzed using SPSS 19 software.

3. Results

Normality test of Erythrocyte data, hematocrit, Hemoglobin and VO2 max have normal distribution and same variance data, thus data analysis done on T-test is not in pair. The results of the research show that the average of Haematology parameter consisted of Erythrocyte, Hematocrit and Hemoglobin level on woman with menstrual period as listed in Table 1. Based on the results in Table 1, it can be seen that the average Erythrocite in the P1 group (4.30 ± 0.05 1012 / l) is lower than P2 group (4.61 ± 0.04 1012 / l). The data analysis obtained value p= 0.000 which means there is a significant difference of Erythrocytelevel on menstrual woman who get iron and non-iron. Table 1 shows that the average hematocrit in group P1 (37.03 ± 0.05%) is lower than group P2 (40.93 ± 0.37%). The data analysis obtained p = 0,000, which means that there is significant difference of Hematocrit level on menstrual woman who get iron and non-iron. Table 1 shows that the average Hemoglobin in group P1(12.22 ± 0.41g / dl) is lower than group P2(14.17 ± 0.24g / dl). The data analysis obtained p = 0,000, which means that there is the difference of Hemoglobin level onmenstrual woman who get iron and non- iron. The results obtained the average difference VO2 max on menstrual women as listed in table 2.

Table 1. Average difference VO_2 max on menstrual woman

Treatment group	**Mean ± SD**	**P**
P1	38,61± 0,26	0,000*
P2	40,62± 0,53	

Notet : (P1) control; (P2) receive iron

Considering on the results of Table 2, it can be seen that the mean VO2 max in group P1 (38.61 ± 0.26 ml / kg Kg / mnt) was lower than group P2(40.62 ± 0.53ml / Kg BW / min) . Data analysis obtained p value = 0.000 which means there is significant difference vO2 max onmenstrual woman who receive iron.

4. Discussion

This research was conducted to find out the effect of iron on VO2 Max and some Haematologyparameters in women who experience menstruation. Thisresearch found that iron has an effect on the increase of erythrocytes, hematocrit, hemoglobin level and VO2 Max in Sports Science students who experience menstruation. This research is in line with previous research by Pasricha et.al (2014) which states that iron supplements can significantly increase VO2 max and performance of menstrual women.This explains that iron can prevent and treat iron deficiency on women of childbearing age (reproductive age).

Women of childbearing age are at high risk of iron deficiency and iron deficiency anemia due to blood coming out during menstruation. Iron deficiency takes place when blood loss causes reduced hemoglobin and iron levels in the body while generally consumed foods do not contain adequate iron, especially for women with menstrual periods (DellaValle, 2001).

Iron is essential for many functions related to physical activity and exercise. WHO recommends the distribution of iron supplements to all women where the prevalence of anemia (in women) exceeds 20% (WHO, 2011). The International Olympic Committee also recommends female athletes to screen for iron deficiency in order to improve performance (Ljungqvist et al., 2009). VO2 Max called as maximum aerobic capacity is the body's ability to consume oxygen maximally per minute. High and

IOP Conf. Series: Journal of Physics: Conf. Series **970** (2018) 012014 doi:10.1088/1742-6596/970/1/012014

low VO2 Max on a person is influenced by several factors such as: the ability of the heart to pump blood throughout the body; the capacity of the lung that serve to take oxygen from the outside air; the quality of blood (hemoglobin) that serves to bind oxygen and is carried throughout the body, the blood vessels that channel blood throughout the body and the ability of the skeletal muscle to use oxygen for metabolism (oxidation) so as to produce a lot of energy to support the old physical activity (endurance) (Foss and Keteyian, 2006).

In addition to supporting the aerobic metabolism process, VO2 Max is also required for recovery process. After a long and heavy activity, the supply of ATP in the active muscle cell is so low that it is insufficient to generate energy on muscle contraction, therefore ATP recepture process requires energy derived from KreatinPosphat, glucose through anaerobic glycolysis process and oxidation process glucose and fat. The oxidation process occurs in mitochondria and this requires oxygen taken by iron-binding hemoglobin (ferritin). thus the sufficient amount of oxygen is needed for sports that require endurance or in other words a person must have a good VO2 Max.

5. Conclusion

Thesupplyingof iron to menstrual woman affect the increase of erythrocytes, hematocrit, hemoglobinlevel and VO2 Max

References

[1] Clement DB, Asmundson RC 1982. Nutritional intake and hematological parameters in endurance runners. Physic Sport Med. Vol. 10:37- 43.

[2] DellaValle DM. 2001. Iron supplementation for female athletes: effects on ironstatus and performance outcomes. Curr Sports Med Rep. 2013;12:234–9.

[3] Foss ML, Keteyian, SJ 2006, *Physiological basis for exercise and sport*, Mc.Graw- Hill Companies, New York, pp. 59-64.

[4] Haas J, Brownlie T 2001. *Iron deficiency anemia and reduced work capacity: a critical review of the research to determine a causal relationship*. J Nutr. Vol.131:676S–90S.

[5] Hinton P, Giordano C, Brownlie T, Haas J 2000. *Iron supplementation improves endurance after training in iron-depleted, non-anemic women*. J Appl Physiol. Vol. 88:1103–11.

[6] Ljungqvist A, Jenoure P, Engebretsen L, Alonso JM, Bahr R, Clough A, Bondt GD, Dvorak J, Maloley R, Matheson G, et al. 2009. The International Olympic Committee (IOC) Consensus Statement on periodic health evaluation of elite athletes March 2009. Br J Sports Med. Vol. 43:631–43.

[7] Newhouse II. Clement DB. 1988. Iron status in athletes. Sports Med. Vol. 5: 337-52.

[8] Pasricha SR, Low M, Thompson J, Farrell A, De-Regil LM 2014. Iron Supplementation Benefits Physical Performance in Women of Reproductive Age: A Systematic Review and Meta-Analysis [1–3]. *The Journal of Nutrition.*

[9] Roseann M L, Connie M Weaver, Darlene A S, Sujaiha Rajaram, Berdine Marlin, and Christopher L M (1992). Iron status in exercising women: the effect of oral iron therapy vs increased consumption of muscle foods. *Am. J. Clin. Nutr.*56:1049-55.

[10] Weaver C. M., Rajaram S (1992) *Exercise and Iron Status*. American Institute of Nutrition. *J Nutr* 1992; 122: 782-7.

[11] World Health Organization 2011. Guideline: Intermittent iron and folic acid supplementation in menstruating women. Geneva, Switzerland: World Health Organization.

[12] Zhu, Y. I., and J. D. Haas (1997) Iron depletion without anemia and physical performance in young women. *Am. J. Clin. Nutr.* 66: 334–341, 1997.

The Simulation Computer Based Learning (SCBL) for Short Circuit Multi Machine Power System Analysis

Rahmaniar[1a], **Maharani Putri**[2]
[1,2]/Power Engineering Research Group Department Of Electrical Enginering
Faculty Of Engineering-Universitas Pembangunan Panca Budi (UNPAB)

[a]Correspondingauthors: niar4n01@yahoo.com

Abstract-Strengthening Competitiveness of human resources become the reply of college as a conductor of high fomal education. Electrical Engineering Program UNPAB (Prodi TE UNPAB) as one of the department of electrical engineering that manages the field of electrical engineering expertise has a very important part in preparing human resources (HR), Which is required by where graduates are produced by DE UNPAB, Is expected to be able to compete globally, especially related to the implementation of Asean Economic Community (AEC) which requires the active participation of graduates with competence and quality of human resource competitiveness. Preparation of HR formation Competitive is done with the various strategies contained in the Seven (7) Higher Education Standard, one part of which is the implementation of teaching and learning process in Electrical system analysis with short circuit analysis (SCA) This course is a course The core of which is the basis for the competencies of other subjects in the advanced semester at Development of Computer Based Learning model (CBL) is done in the learning of interference analysis of multi-machine short circuit which includes: (a) Short-circuit One phase, (B) Two-phase Short Circuit Disruption, (c) Ground Short Circuit Disruption, (d) Short Circuit Disruption One Ground Floor Development of CBL learning model for Electrical System Analysis course provides space for students to be more active In learning in solving complex (complicated) problems, so it is thrilling Ilkan flexibility of student learning how to actively solve the problem of short-circuit analysis and to form the active participation of students in learning (Student Center Learning, in the course of electrical power system analysis.
Key Word. Active Learning CBL, Short Curcuit Anaylsis

1. Introduction

The electric power system that generally consists of a channel and load generator is a system that processes the empowerment and use of electrical energy.The need for electrical energy is a primary priority in supporting development activities in various aspects of life [1].So Conducted several studies and studies that are analyzed to improve reliability in the process of empowerment and use of electrical energy.In operation, the power system can not be separated from interference, be it interference that is temporary or permanent disturbance. Interference in the power system consists of symmetrical disruption, asymmetric interference and open conductor disruption [2].

IOP Conf. Series: Journal of Physics: Conf. Series **970** (2018) 012015 doi:10.1088/1742-6596/970/1/012015

The complex power system implies that the system has multiple machines in the service of the load, so special handling is required in providing customer service continuity by maintaining, effectively handling disturbance conditions so that the distribution of electrical energy does not experience significant constraints .From this it is necessary to conduct a series of analysis of electric power system, one of which is a brief short-circuit analysis using a practical computer.

The learning model (CBL) is a form of computer-based learning model. Computers are used as a medium / tool to deliver materials and teaching materials in the learning process to be able to attract the attention of learners.The teaching materials to be delivered are developed using a learning development model that can be used to create computer-based learning.In the teaching materials, it has included material, questions, and feedback or tasks presented in a single program package.CBL learning model is a teaching and learning system that uses computer equipments as a tool of help together with its knowledge base (knowledge base), is the development of integrated information technology ie communication (interactive), audio, video, image appearance Packed with multimedia designations.WithBenefits of using CBL learning model, among others: (A) Able to reduce training costs, (b) Flexibility of time, (c) Flexibility of learning speed, (d) Standardization of learning, (e) Effectiveness of learning, (f) Can store student data, lessons and ongoing learning process [2] .

2. The Computer Based Learning (CBL) For Short Circuit Analysis.

The design of instructional media designed to produce an interactive learning media for protection system applied to industrial electrical system, In the sample paper has presented plot of relay coordination response curve for industrial system grid system 3 bus By using this package students get simulation experience as well as knowledge about Industrial applications. The created package has been able to complete the course activity and as a teaching tool For lecturers and students. Computing The program is designed to help students avoid the boredom to calculate the industrial security system by replacing the method of calculation with a simulation dish. This approach can provide a very important stimulus in building self-learning concepts for students [3]

The simulated model built asks and directs them (the students) to provide space to discuss the results of short-circuit analysis on the outcomes of the project, (3) Report on simulation results in practical practical laboratory-models of their PSP simulation experiments in learning. And observing And measure Tingat motive Studying students from the PSP-based model of computer simulation experiments. Network modeling, power flow and short-circuit current analysis and Over Current Protection (OC) and coordination of protective relays have been presented using computer-assisted software in ETAP. In addition, students can visualize the results of observing domain errors with OC, distance and differential protection in software PSCAD / EMTDC. Protection for transformer equipment from Over Current (OC), simulation Over voltage and phase imbalance for induction motor device protection have been tested in practical experiment learning with the help of LabVolt electrical module. Experimental results from Study paper with title Enhancing Student Experience Learning System Power Protection using Computer based simulation with Practical Experiment development can be concluded learning design with computer help facilitated simulation using ETAP and PSCAD / EMTDC software for student students at Curtin University, Australia Conducted in the full semester of learning gets feedback given to the student With an indicator of student interest in Power System Protection (PSP) learning [4,5]

IOP Conf. Series: Journal of Physics: Conf. Series **970** (2018) 012015 doi:10.1088/1742-6596/970/1/012015

3. CBL Design in teching-learning

CBL-based interactive learning environment for short circuit analysis of electrical power system is realized by four stages as shown in Figure 1. Other (1) Preparation of learning planning, by arranging the learning material structure in the form of horizontal matrix containing the contents about the expected competence, teaching materials Which will be delivered as well as indictment of successful learning. The arrangement of this lesson plan will be socialized to the group of lecturers in the field of electrical power system expertise which becomes The same standard of learning plan among lecturers In a field relevant to the electrical systems analysis course (2). Conduct construction of the matrix Z model matrix arrangement of a multi-machine interconnection system which will be arranged in a symmetric short circuit settlement program and the asymmetric short circuit. (3) The results of step 2 (2) further develop CBL-based learning modules and compose CBL-based SCL job-sheets used by learners in learning Solve SCC problem. (4). Completion of cases with data obtained in real terms. To complete CBL learning with project based learning content in the sense that learners are directed to conduct research related to SCC conducted at the time of learning. Development of this learning model Foster interest in learners by participating actively in the learning process of teaching SCC

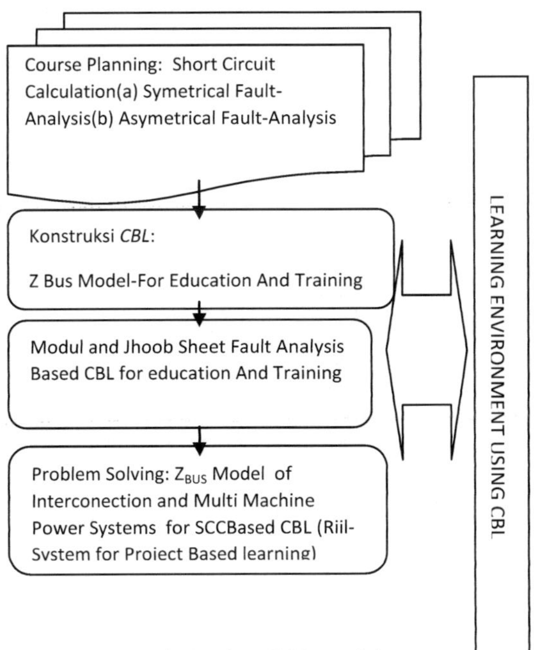

Figure 1. Design CBL model

Teaching and Learning Process of short circuit interaction analysis of power systems developed by innovation Lecturing system that is not only done in the lecture room, but developed by implementing the practical learning in the computer laboratory, basically the course is the subject of theoretical studies. However, the development of an interactive CD is completed. CBL learning model for schematic SCC study is shown in fig.1

For Practical Computer Based Learning Device [6], illustration of power system diagram is given in figure 2

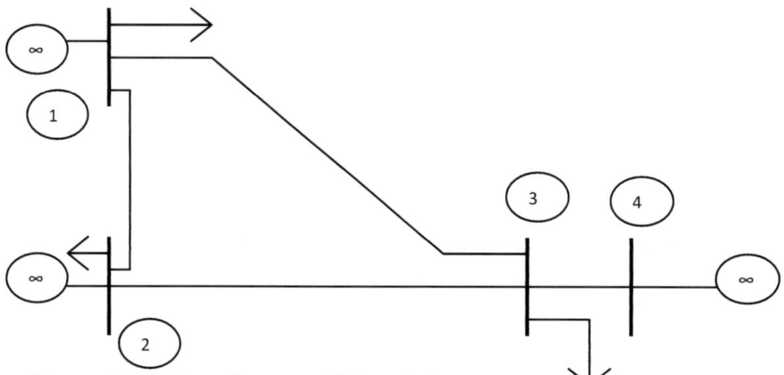

Figure 2. One line diagram of Electric Power system

The line diagram of the electric power system in Fig.. 2 will be formed in the reactance diagram, as shown in Fig. 3 below:

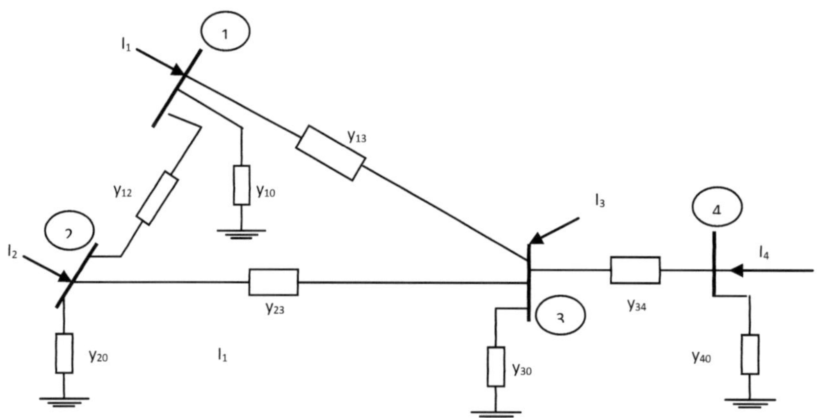

Figure 3. Admittance Diagram

From the picture of the admittance diagram figure (3) is prepared the equation:

$$I_1 = V_1\, y_{10} + (V_1 - V_2)\, y_{12} + (V_1 - V_3)\, y_{13} \quad (1)$$

$$I_2 = V_2\, y_{20} + (V_2 - V_1)\, y_{12} + (V_2 - V_3)\, y_{23} \quad (2)$$

$$I_3 = V_3\, y_{30} + (V_3 - V_1)\, y_{13} + (V_3 - V_4)\, y_{34} + (V_3 - V_2)\, y_{23} \quad (3)$$

$$I_4 = V_4\, y_{40} + (V_4 - V_3)\, y_{34} \quad (4)$$

Equations (1) through (4) can be simplified to be

$$I_1 = (y_{10} + y_{12} + y_{13})\, V_1 \;\; - y_{12}\, V_2 \;\; y_{13}\, V_3 \;\; +0{,}0\, V_4 \quad (5)$$

$$I_2 = \;\; - y_{13}\, V_1 \;\; (y_{20} + y_{12} + y_{13})\, V_2 \;\; y_{23}\, V_3 \;\; +0{,}0\, V_4 \quad (6)$$

$$I_3 = \; - y_{13}\, V_1 - y_{13}\, V_2 + (y_{30} + y_{12} + y_{23} + y_{34}\;)V_3 + y_{34}\, V_4) \quad (7)$$

$$I_1 = \;\; 0{,}0V_1 + 0{,}0\, V_2\, y_{34}\, V_3 \;\; +(y_{40} + y_{34})\, V_4 \quad (8)$$

Can be arranged into Y Bus matrix and generate bus impedance matrix: ZBUS = $YBUS^{-1}$

Recognize the massive short-circuit symmetrical short-circuit impedance through a series of conventional analyzes and computer-based testing to derive practical and efficient ways of determining the value of short-circuit power circuit breakdowns.

4. Case Study (Simulated System Model)

System 10 bus as shown below with dataas follows

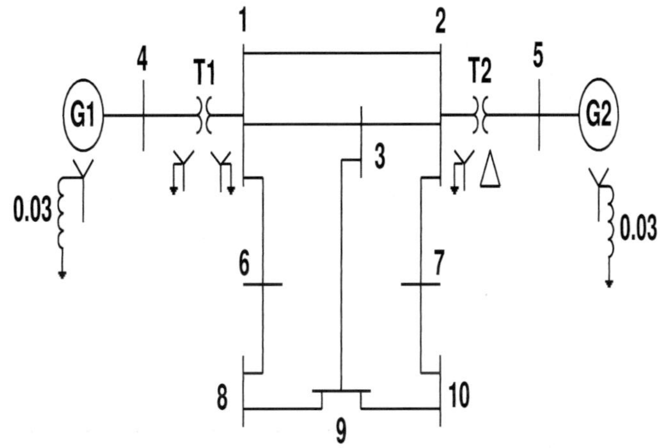

Figure 4. 10-bus interconnection power system

Table 1. Data 10-Bus Power System

ITEM	MVA RATING	VOLTAGE RATING	X_1	X_2	X_0
G1	100	25 kV	0.20	0.20	0.05
G2	100	13.8 kV	0.20	0.20	0.05
T1	100	25/230 kV	0.05	0.05	0.05
T2	100	13.8/230 kV	0.05	0.05	0.05
1-2	100	230 kV	0.10	0.10	0.30
2-3	100	230 kV	0.10	0.10	0.30
1-3	100	230 kV	0.10	0.10	0.30
1-6	100	230 kV	0.40	0.40	0.80
2-7	100	230 kV	0.40	0.40	0.80
3-9	100	230 kV	0.40	0.40	0.90
6-8	100	230 kV	0.10	0.10	0.10
8-9	100	230 kV	0.40	0.40	0.60
7-10	100	230 kV	0.40	0.40	0.90
9-10	100	230 kV	0.40	0.40	0.80

Matrix Z modeling bus is used to input the reactance data on M. Matlab to obtain Positive sequence, Negative sequence and Zero sequence, Impedance Matrix with equation [7,8]

$$Z_{rel} = [Y_{rel}]^{-1}$$

IOP Conf. Series: Journal of Physics: Conf. Series **970** (2018) 012015 doi:10.1088/1742-6596/970/1/012015

5. Result simulation

SCC Simulation for 11 Bus System Data Input on M. Matlab Files and Enter Faulted Bus No. -> 8 Enter Fault Impedance Zf = R + j*X in complex form (for bolted fault enter 0). Zf = 0.1, Balanced three-phase fault at bus No. 8. Total fault current = 6.3869 per unit Bus Voltages during fault in per unit:

Bus No.	Voltage Magnitude	Angle degrees
1	0.8187	-11.5749
2	0.7801	-15.2615
3	0.7466	-19.1233
4	0.7311	-21.1824
5	0.7033	-25.5573
6	0.6910	-27.8702
7	0.7356	-20.5678
8	0.6387	-50.3057
9	0.6570	-36.7585
10	0.6449	-42.3209
11	0.6633	-34.6346

Line currents for fault at bus No. 8

From Bus	To Bus	Current Magnitude	Angle degrees
G	1	1.2861	-50.3057
1	2	1.2861	-50.3057
2	3	0.6141	-50.3057
2	11	0.6719	-50.3057
3	4	0.6141	-50.3057
4	5	0.6141	-50.3057
5	6	0.6141	-50.3057
6	8	2.6371	-50.3057
G	7	2.0230	-50.3057
7	6	2.0230	-50.3057
8	F	6.3869	-50.3057
G	9	3.0779	-50.3057
9	8	3.0779	-50.3057
10	8	0.2240	-50.3057
11	8	0.4480	-50.3057
11	10	0.2240	-50.3057

6. Conclusion

Based on the research results can be concluded as follows: The electrical system model is formed into the R impedance matrix R equation using M. Matlab facilitates practical analysis of the case of short circuit of electric power system Simulation and analysis results obtained short circuit interruption on bus 8 for the system dialiaaliais respectively: Ifault 3 phase = 6.3869 pu, Ifault for one Phase = 8,5011 pu, Ifault ; Between phases = 3,87 pu Preparation of computer-assisted short-circuit analysis tools in learning (Computer Based Learning) in practical, efficient and fast can be utilized by students of electrical engineering program UNPAB in electrical power system analysis analysis for material Short Circuit Analysis.

Reference

[1] Padiyar, K.R., 1996, ***Power System Dynamis Stability and Control***, pp. 191 – 268, John Wiley & Sons , Singapore.

[2] Moh. E El Hawary "*Electrical* ***Power System*** " IEEE Press. 1983

[3] L. L. La, "Computer Assisted Learning In Power System *Relaying*," *IEEE,* vol. 38 no 3, pp. 217-222, August 1995.

[4]. M. M. Farhad Shahnia, ***"Improving the Learning Experience of Power System Protection Students using Computer-based Simulations and Practical Experiments "*** *AUPEC,* 28 September - 1 October 2014.

[5]. Joong Rin-shin" ***A Windows-Based Interactive and Grafic And Package for Educatión and Training Of Power System Analysis And Operation"*** IEEE Transaction on Power System Vol 14 No. 4 , 1999

[6] Stevenson, W.D., dan Idris,K.,1990, *Analisis Sistem Tenaga Listrik*, pp. 349-351, Erlangga, Jakarta.

[7]. H.F Wang *"A Unified Model for the Analysis of FACTS devices in Damping Power Oscilations-PART I : Single Machnie Infinite Bus Powers Systems, "* IEEE Transaction On Power Delivery, Vol 12 No. 2 pp 941-946, April 1997

[8] P.M. Anderson , A.A Fouad *" Power System Kontrol and stability"* The Iowa State university Press, © 1977

IOP Conf. Series: Journal of Physics: Conf. Series **970** (2018) 012016 doi:10.1088/1742-6596/970/1/012016

PTR, PCR and Energy Resolution Study of GAGG:Ce Scintillator

Pruittipol Limkitjaroenporn[1,2a], Wiraporn Hongtong[1,2], Hong Joo Kim[3] and Jakrapong Kaewkhao[1,2]

[1]Physics Program, Faculty of Science and Technology, Nakhon Pathom Rajabhat University,Nakhon Pathom, 73000, Thailand,
[2]Center of Excellence in Glass Technology and Materials Science (CEGM), Nakhon Pathom Rajabhat University, Nakhon Pathom, 73000, Thailand,
[3]Department of Physics, Kyungpook National University, Daegu 702-701, Korea

Corresponding author: golfpo@hotmail.com

Abstract. In this paper, the peak to total ratio (PTR), the peak to Compton ratio (PCR) and the energy resolution of cerium doped gadolinium aluminium gallium garnet (GAGG:Ce) scintillator are measured in the range of energy from 511 keV to 1332 keV using the radioactive source Na-22, Cs-137 and Co-60. The crystal is coupled with the PMT number R1306 and analyzed by the nuclear instrument module (NIM). The results found that the PTR and PCR of GAGG:Ce scintillator decrease with the increasing of energy. The results of energy resolution show the trend is decrease with the increasing of energy which corresponding to the higher energy resolution at higher energy. Moreover the energy resolution found to be linearly with .

1. Introduction

Normally, the photo detectors coupled with crystal scintillators instruments [1] are used to count of high-energy photons and particles. Scintillator materials are important in many applications such as X-ray computed tomography (CT), positron emission tomography (PET), Single-photon emission computed tomography (SPECT) and other medical imaging techniques and special applications in high energy, nuclear, space and medical physics [2,3]. There are several parameters to confirm good scintillators, i.e. the highest light yield, good energy resolution, high effective atomic number, fast scintillation response, chemical stability and capability of large crystal growth [4].

The researchers were interested in the cerium doped materials such as silicates, LSO:Ce or LYSO:Ce because of the best combination of parameter factors. They have been developed as promising scintillators for PET due to high density, high effective atomic number, fast decay time and high light yield [5,6]. Also cerium-activated lanthanum halides, such as $LaBr_3$:Ce and $LaCl_3$:Ce are reveal very good energy resolution, fast emission and excellent temperature and linear characteristics [7,8]. Oxide materials based on the garnet structure are appropriate choice for applications as scintillators because of high developed technology and easy doping by rare-earth elements for laser hosts and other applications.

IOP Conf. Series: Journal of Physics: Conf. Series **970** (2018) 012016 doi:10.1088/1742-6596/970/1/012016

Recently, gadolinium aluminium gallium garnet (GAGG:Ce) single crystal is grown by the micropull-down method and the results show that the light yield (LY) is 42,000 photons/MeV and energy resolution of 8.3% at photon energy 662 keV which is measured with the avalanche photodiode (APD). The single crystals can be grown by the Czochralski method which show the highest of LY is 46,000 photons/MeV and an emission peak at 530 nm and the range of good energy resolution is 4.9–5.5% at 662 keV [4]. GAGG:Ce is a new single crystal scintillator with several properties that make it interesting for applications such as gamma spectroscopy. GAGG:Ce has high density is 6.5 g/cm^3, similar to LSO, YAG or LuAG, which are generally used for high energy gamma-ray measurements, high gamma-ray absorption coefficient [9].

In this paper, the investigation of the peak to total ratio (PTR), the peak to Compton ratio (PCR) and the energy resolution of GAGG:Ce scintillator were measured in the range of energy from 511 keV to 1332 keV using the radioactive source Na-22, Cs-137 and Co-60.

2. Theory

2.1. Energy Resolution (R)

The Energy resolution (R) is the ability of the detector to determine accurately the energy of the incoming radiation. Energy resolution was calculated from the full width at half maximum (FWHM) of electron energy in the range of energy range from 511 keV to 1332 keV and it can be calculated by the following relation [10]:

$$R = \frac{FWHM}{E_0} \times 100\%, \tag{1}$$

where E_0 is the electron energy peak.

2.2. Peak to total ratio (PTR)

Peak to total ratio (PTR) is the relation between the area of photopeak and the total area in the spectrum. It can be calculated by this formula [11].

$$\text{PTR} = \frac{\text{The area of the photopeak}}{\text{The total area in the spectrum}} \tag{2}$$

2.3. Peak to Compton ratio (PCR)

Peak to Compton ratio (PCR) is the ratio between the height of the photopeak and that of the Compton continuum, which is defined for source of the gamma ray spectrum as the ratio between the height of the photopeak and the average height of the Compton peak [12].

$$\text{PCR} = \frac{\text{The height of the photopeak}}{\text{The average height of the Compton peak}} \tag{3}$$

3. Experimental

Experimental setup shown in Figure 1. The study of cerium doped gadolinium aluminium gallium garnet (GAGG:Ce) scintillator with a size of $10 \times 10 \times 0.8$ mm^3 was measured with a photomultiplier tube (PMT) number R1306, the Hamamatsu Photonics was a manufacturer. Using silicone grease for was coupled between scintillator and PMT, and wrapped in several layers of white teflon tape and wrapped again in black tape and then that covered with aluminium housing. The signal output from the PMT anode was passing through a canberra 2007B preamplifier and was sent to canberra 2022 an amplifier. These experimental set up was using nuclear instrument module (NIM) and shaping time of the spectroscopy amplifier were controlled at 0.5 μs for all measurements. The energy spectra

IOP Conf. Series: Journal of Physics: Conf. Series **970** (2018) 012016 doi:10.1088/1742-6596/970/1/012016

analyzed with software of canberra MCA to record data [13]. In this work the study of the peak to total ratio (PTR), the peak to Compton ratio (PCR) and the energy resolution of GAGG:Ce scintillator were measured the radioactive source Na-22, Cs-137 and Co-60 in the energy range of 511 keV to 1332 keV.

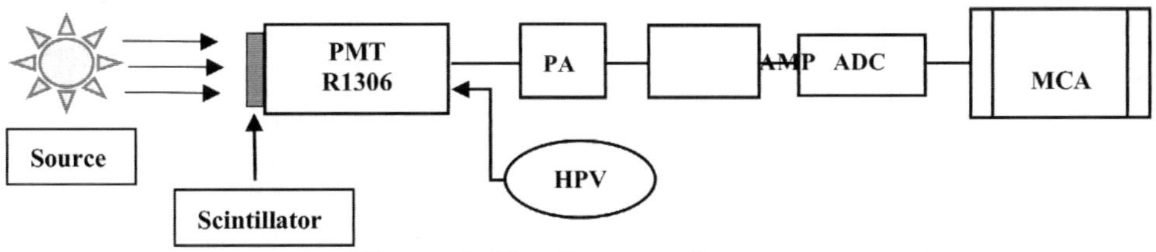

Figure 1: The diagram of gamma ray spectroscopy

4. Result and Discussion

4.1. Energy Resolution

The energy resolution of the GAGG:Ce scintillator was determined using Eq. (1) by the different radioactive source in the energy range from 511 keV to 1332 keV and shown in Figure 2. The results of the energy resolution of GAGG:Ce scintillator increase proportional to the decreasing of energy from 511 keV to 1332 keV.

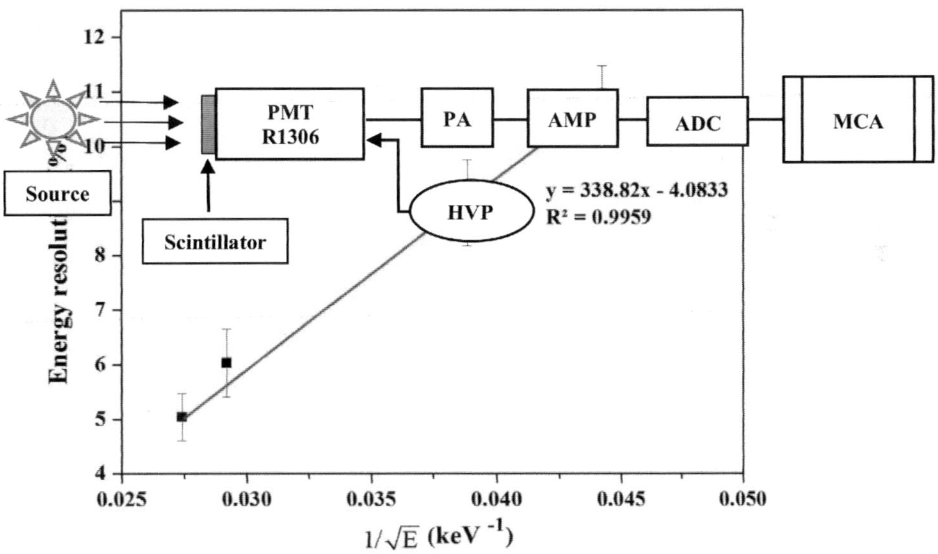

Figure 2: Energy resolution of GAGG:Ce in the energy range 511 keV to 1332 keV

4.2. Peak to total ratio and Peak to Compton ratio

The peak to total ratio (PTR) and peak to Compton ratio (PCR) are determined by using Eq. (2) and Eq. (3) of GAGG:Ce in the energy range from 511 keV to 1332 keV. In Figure 3 presents the results of the peak to total ratio of GAGG:Ce in the energy range from 511 keV to 1332 keV and show that the values decrease from 1.096 to 0.196 with the increasing of the energy from 511 keV to 1332 keV.

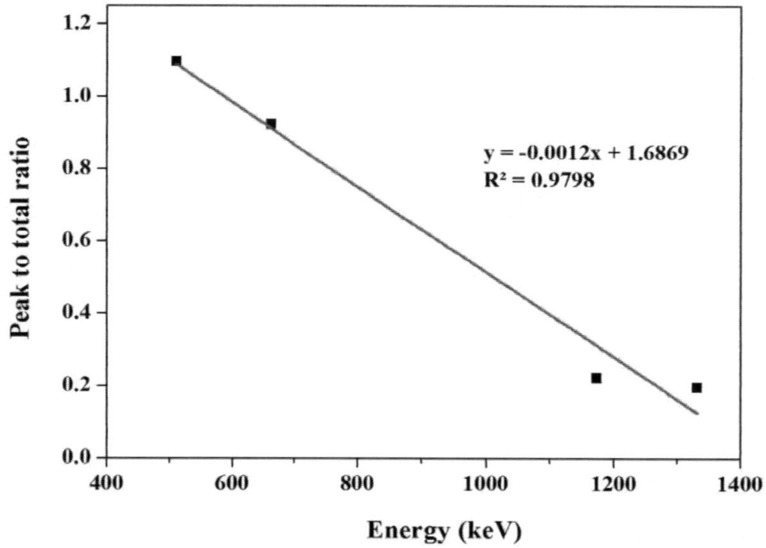

Figure 3: Peak to total ratio of GAGG:Ce in the energy range from 511 keV to 1332 keV

The peak to Compton ratio for GAGG:Ce scintillator is shown in Figure 4. The results of peak to Compton ratio decrease from 2.98 to 0.60 with the increasing of the energy range from 511 keV to 1332 keV and the trend is also the same as the peak to total ratio.

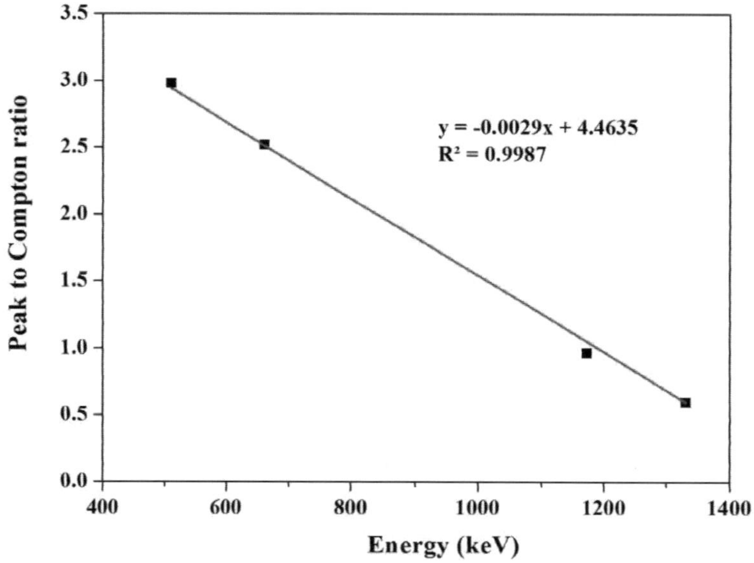

Figure 4: Peak to Compton ratio of GAGG:Ce in the energy range from 511 keV to 1332 keV

5. Conclusion

In this paper, the peak to total ratio (PTR), the peak to Compton ratio (PCR) and the energy resolution of cerium doped gadolinium aluminium gallium garnet (GAGG:Ce) scintillator were measured in the energy range from 511 keV to 1332 keV using the radioactive source Na-22, Cs-137 and Co-60. The trend of energy resolution is inversely proportional to the square root of the energy and it increased with the decreasing of the energy. The results of the peak to the total ratio and the peak to the Compton ratio of GAGG:Ce scintillator decreased with the increasing of the energy.

Acknowledgements
The authors would like thank to Department of Physics, Faculty of Science and would like to acknowledge the National Research Council of Thailand (NRCT) and Nakhon Pathom Rajabhat University (NPRU) for financial supports.

References
[1] O. Sakthong, W.Chewpraditkul, C. Wanarak, K. Kamada, A. Yoshikawa, P. Prusa and M. Nikl, "Scintillation properties of $Gd_3Al_2Ga_3O_{12}:Ce^{3+}$ single crystal scintillators," *Nucl. Instrum. Methods Phys. Res., Sect. A.*, vol. 751, pp. 1-5, 2014.

[2] J. Krimmer, J.-L.Ley, C. Abellan, J.P. Cachemiche, L. Caponetto, X. Chen, M. Dahoumane, D. Dauvergne, N. Freud, B. Joly, D. Lambert and L. Lestand, "Development of a Compton camera for medical applications based on silicon strip and scintillation detectors," *Nucl. Instrum. Methods Phys. Res., Sect. A.*, Vol. 787, pp. 98–101, 2015.

[3] M. Kiunke, C. Stritt, R. Schielein, F. Sukowski, A. Hölzing, S. Zabler, J. Hofmann, A. Flisch, S. Kasperl, U. Sennhauser, R. Hanke, ROSI and GEANT4, "A comparison in the context of high energy X-ray physics," *Nucl. Instrum. Methods Phys. Res., Sect. B.,* vol. 377, pp. 50–58, 2016.

[4] J. Iwanowska, L. Swiderski, T. Szczesniak, P. Sibczynski, M. Moszynski , M. Grodzicka , K. Kamada, K. Tsutsumi, Y. Usuki, T. Yanagida, A. Yoshikawa," Performance of cerium-doped $Gd_3Al_2Ga_3O_{12}$ (GAGG:Ce) scintillator in gamma-ray spectrometry," *Nucl. Instrum. Methods Phys. Res., Sect. A.,* vol. 712, pp. 55-70, 2015.

[5] F. Yang, R. Mao, L. Zhang, and R.Y. Zhu, "Characterization of three LYSO crystal batches," *Nucl. Instrum. Methods Phys. Res., Sect. A.,* vol. 784, pp. 105-110, 2015.

[6] D. W. Cooke, K. J. McClellan, B. L. Bennett, J. M. Roper, M. T. Whittaker, R. E. Muenchausen, and R. C. Sze, "Crystal growth and optical characterization of cerium-doped Lu 1.8 Y 0.2 SiO 5," *J. Appl. Phys.,*vol, 88, no. 12, pp. 7360-7362, 2000.

[7] E. V. D. van Loef, P. Dorenbos, C. W. E. van Eijk, K. Krämer, and H. U. Güdel, "High-energy-resolution scintillator: Ce^{3+} activated $LaBr_3$," *Appl. Phys. Lett.,* vol. 79, no. 10, pp. 1573-1575, 2001.

[8] E. V. D. van Loef, P. Dorenbos, C. W. E. van Eijk, Member, IEEE, K. Krämer, and H. U. Güdel, "Scintillation Properties of $LaCl_3:Ce^{3+}$ Crystals: Fast, Efficient, and High-Energy Resolution Scintillators," in *Proceedings of the IEEE transactions on nuclear science*, 2001, vol. 48, pp. 341-345.

[9] K. Kamada, T. Yanagida, J. Pejchal, M. Nikl, T. Endo, K. Tsutsumi, Y. Fujimoto, A. Fukabori and . Yoshikawa, "Crystal Growth and Scintillation Properties of Ce Doped Single Crystals," in *Proceedings of the IEEE transactions on nuclear science*, vol. 11, pp. 4484 –4490, 2011.

[10] J. Kaewkhao, P. Limkitjaroenporn, W. Chaiphaksa and H.J. Kim, "Non-proportionality study of $CaMoO_4$ and GAGG:Ce scintillation crystals using Compton coincidence technique," *Appl. Radiat. Isot.*, vol. 115, pp. 221–226, 2016.

[11] H. Leutz, G.SchulzL and V. Gelderen, "Peak/total-ratios for NaI(Tl)-crystals," *Nucl. Instr. Meth.* vol. 40 pp. 257-260, 1966.

[12] Y. M. Tsipenyuk, "Physical methods, instruments and measurements," oxford, United Kingdom, 2009, pp. 81.

[13] W. Chaiphaksa1, N. Boonyopakorn, J. Kaewkhao, P. Limkitjaroenporn, "Light Yield Non-proportionality and Energy Resolution of BGO and CsI(Tl) Scintillation Crystals," *Advanced Materials Research,* vol. 979, pp. 285-288, 2014.

IOP Conf. Series: Journal of Physics: Conf. Series **970** (2018) 012017 doi:10.1088/1742-6596/970/1/012017

Physical Characteristics of Laboratory Tested Concrete as a Substituion of Gravel on Normal Concrete

Ronald Butar-butar[1], Suhairiani[2], Kinanti Wijaya[3], Nono Sebayang[4]
[1,2,3,4]Department of Engineering Building, State University of Medan, Medan, Indonesia

*Corresponding Author: *suhairiani@unimed.ac.id*

Abstract - Concrete technology is highly potential in the field of construction for structural and non-structural construction. The amount uses of this concrete material raise the problem of solid waste in the form of concrete remaining test results in the laboratory. This waste is usually just discarded and not economically valuable. In solving the problem, this experiment was made new materials by using recycle material in the form of recycled aggregate which aims to find out the strength characteristics of the used concrete as a gravel substitution material on the normal concrete and obtain the value of the substitution composition of gravel and used concrete that can achieve the strength of concrete according to the standard. Testing of concrete characteristic is one of the requirements before starting the concrete mixture. This test using SNI method (Indonesian National Standard) with variation of comparison (used concrete : gravel) were 15: 85%, 25: 75%, 35:65%, 50:50 %, 75: 25%. The results of physical tests obtained the mud content value of the mixture gravel and used concrete is 0.03 larger than the standard of SNI 03-4142-1996 that is equal to 1.03%. so the need watering or soaking before use. The water content test results show an increase in the water content value if the composition of the used concrete increases. While the specific gravity value for variation 15: 85% until 35: 65% fulfilled the requirements of SNI 03-1969-1990. the other variasion show the specifics gravity value included on the type of light materials.
Keywords: *Characteristic; Substitution gravel; Used concrete; Waste*

1. Introduction

The main building materials that are widely used in the construction of buildings are concrete, such as buildings, bridges, docks, houses, and others. The development of concrete technology has great potential in the field of construction, both for structural and non-structural construction. The use of concrete materials in buildings is not only done by Indonesia but throughout the country. The amount used of this concrete material raises the problem of solid waste in the form of concrete residual strength test results which is a requirement in building a building and concrete residual collapse of the building that will do the renovation. This concrete waste is usually only discarded and not economically valuable.

In solving the problem, a new concrete test is made using recycled aggregate material. Based on Suharwanto's [1] research results, recycled aggregates contain 25-45% mortar for coarse aggregates, and 70-100% for fine aggregates. This affects the water uptake that will occur in the new concrete.

The recycled aggregate has a tensile strength value and its poisson ratio is relatively the same as the natural aggregate concrete. As well as the typical values of reconstituted recycled aggregate concrete elements also have a performance and collapse pattern that is relatively equal to the typical value of natural aggregate concrete elements, only slightly lower ductility values [2]. The recycled aggregate composition affects the shrinkage of the concrete. The more recycled aggregate composition the shrinkage value on the concrete is greater. To overcome this problem, the mixing of homogeneous concrete was done by adding subtle patio materials to the mixture [3].

This research was done to utilizing used concrete as a normal concrete material so that it can reduce the price of concrete. The use of recycled materials is not known in general. However, using recycled aggregates can reduce the price of each volume of concrete, since the ingredients are obtained from waste that is relatively lower than the natural aggregate. The low price of concrete will affect the ability of the community to build residential houses to be permanent, so that the rate and expansion of the development can be increased up to the remote villages.

Aggregates are natural mineral granules that serve as fillers in mortar or concrete mixtures. This aggregate occupies approximately 78% of the volume of mortar or concrete. Although only as fillers but aggregates greatly affect the properties of mortar or concrete so that aggregate selection is an important part in the manufacture of mortar or concrete. As with the alternative use of burned concrete fractions as a coarse aggregate, because the condition at this time aggregate began to decrease and the price soared. This kind of thing is experienced by some areas that are difficult to get material for the building, because some areas of material resources are forced to close.

Based on SNI-03-6820-2002 [5] there are requirement for gravel in concrete. The larger the fine modulus values of an aggregate means the greater the aggregate grain (the more rough). The fine modulus of sand grains ranges from 1.50 - 3.80. The fine modulus of gravel grains range from 6.0 - 8.0. The fine-grained modulus of the coarse and fine aggregate mixture is 4.0- 6.0. For mixed aggregate depending on the maximum grain size. The fine modulus of the ideal aggregate grain is based on its grain size, namely: (1)The maximum grain size of 40 mm ranges from 5.5 to 6.0, (2)Maximum grain size of 20 mm ranges from 5.0 - 5.5, (3)Maximum grain size of 10 mm ranges from 4.0 to 4.5. Also for mud contain in fine aggregate (sand) no more than 5% and 1% in gravel.

In concrete, coarse aggregate and fine aggregate fills most of the volume of concrete that is between 60%-80% so that the nature and quality of the aggregate influence on the nature and quality of concrete [6]. The function of the use of aggregate in the concrete is to produce a great strength of the concrete, with good gradation then we will get a solid concrete, and controlling *the workability* or properties of the concrete mix workable. With a good aggregate, it will get the concrete easier to work or have a good workability.

The used concrete is an unused concrete material for construction. The used concrete used in this study is an industrial test specimen conducted in several university laboratories in Medan.

2. Reseach Method

This research is a laboratory experiment research. The basic materials in this research are cement, sand, natural gravel, recycled gravel (used concrete), and water. In general, this research is divided into two main phases that will be implemented, that is:

2.1 Preparation

At this stage the research begins by determining the compressive strength to be achieved. Then determine the materials of used concrete to be used as gravel substitution on the normal concrete. The used concrete used is the result of laboratory testing of industrial samples in the form of cylinders 15x30 cm in size (as Fig.1). The used concrete is then destroyed manually with a maximum size of 40 mm.

Figure 1. Laboratory tested concretes

For the natural aggregate used is taken in the batching plan industry. The aggregate used are gravel stone type which is maximal size ¾ ". And the sand material used is the standard sand commonly used for normal concrete mixture. As well as cement materials used are material that meets national quality standards so that for testing the characteristics of cement are no longer needed.

2.2 *Stages of completion*

At this stage, the physical properties of used concrete as pebble substitution in concrete refer to SNI standard. Comparison of natural recycled aggregate material mixtures and natural aggregates (used concrete: gravel) used in this study were 15: 85%, 25: 75%, 35:65%, 50:50%, 75: 25%. Further laboratory data obtained will be analyzed referring to Indonesian National Standard for normal concrete by taking the mean (average) from the test result data. These tests include testing of moisture content, mud content, specific gravity, water absorption, and filter analysis.

3. Result And Discussion

Preliminary results were achieved in this study is preliminary data research, the water content, silt content, specific gravity and absorption of the former test material mixed concrete and gravel. As well as sieving for fine aggregate (sand).

3.1 Coarse Aggregate Water Content

Moisture content is the magnitude of the ratio between the weight of water contained in the aggregate on aggregate in the dry state (SNI 03-1971 - 1990). Research results obtained from testing the water content of the coarse aggregate (gravel + concrete used) are as follows:

Table 1. Data Result of Water content examination

Examination	Unit	Result	SNI Requirements
Water Content			-
a) Fine Aggregate	%	a) 1.43	
b) Coarse Aggregate	%	b) 0.47	
c) Used Concrete	%		
1) 15 : 85		1) 0.95	
2) 25 : 75		2) 1.05	
3) 35 : 65		3) 1.40	
4) 50 : 50		4) 1.87	
5) 75 : 25		5) 2.57	

The test results obtained from the data the greater the percentage of broken stone, the greater the water content is generated, but if the percentage of gravel is greater then the moisture content value obtained would be lower.

3.2 Levels of Mud

Testing sludge levels aims to determine the levels of sludge contained by fine and coarse aggregates in the laboratory (SNI 03-4142-1996). Fine aggregate and coarse aggregate are good for building materials is for fine aggregate that has a sludge levels below 5%, and for coarse aggregate that has a sludge levels below 1%. if the sludge content of coarse and fine aggregate in excess of that set by SNI then sand or gravel should be washed before being used as a building material. From the research results obtained testing the levels of silt to coarse aggregate (gravel + concrete used) are as follows:

Table 2. Levels of Mud test results

Examination	Unit	Result	SNI Requirements
Mud Content			
a) Fine Aggregate	%	a) 3.01	a) < 5
b) Coarse Aggregate	%	b) 1.03	b) < 1
c) Used Concrete	%		c) < 1
1) 15 : 85		1) 1.05	
2) 25 : 75		2) 1.06	
3) 35 : 65		3) 1.11	
4) 50 : 50		4) 1.13	
5) 75 : 25		5) 1.21	

It can be seen from table 2 that the greater the ratio of the used concrete to the mud found more and more, it is proven by the amount of dust dust attached to the former concrete at the time of solving the former concrete so that the limit determined by SNI 03-4142-1996 has passed the threshold is 1%. However, the mud content for fine aggregate is still within the specified threshold limit of less than 5%.

3.3 Coarse Aggregate Specific Gravity

Specific gravity test aims to determine the specific weight of coarse aggregate and its ability to absorb water. The amount of density that is checked is to aggregate in a dry state, a saturated surface-dry weight (Saturated Surface Dry), apparent specific gravity (Apparent). Density used is a dry density surface (SSD), if the terms of its density aggregate divided into three kinds. (1)Lightweight Aggregate, aggregateis an aggregate that has a specific gravity less than 2.0, and is usually used for non-structural concrete; (2)Aggregate normal, aggregate normal is the aggregate that has a specific gravity of 2.5 to 2.7. The result of this specific gravity of concrete is about 2.3 with a compressive strength of 15 MPa to 40 MPa.; (3)Weight aggregate, aggregate is pick a specific gravity of more than 2.8. Concrete produced also has a high density (up to 5.0), which is effective as an X-ray radiation protective.

From the results obtained testing the specific gravity for coarse aggregate (gravel + concrete used) are as follows:

Table 3. Specific Gravity results of aggregate

Examination	Unit	Result	SNI Requirements
Specific Gravity			
a) Fine Aggregate	gr/cm	a) 2.51	a) ≥ 2.5
b) Coarse Aggregate	gr/cm	b) 2.52	b) ≥ 2.5
c) Used Concrete	gr/cm		c) ≥ 2.5
1) 15 : 85		1) 2.53	
2) 25 : 75		2) 2.56	
3) 35 : 65		3) 2.55	
4) 50 : 50		4) 2.49	
5) 75 : 25		5) 2.43	

The data from test results obtained by the average - average specific gravity for coarse aggregate is, the specific gravity value for variation 15: 85% until 35: 65% fulfilled the requirements of SNI 03-1969-1990. the other variasion show the specifics gravity value included on the type of light materials.

3.4 Absorption Coarse Aggregates

Absorption Testsaims to determine the seepage in coarse aggregate (gravel) used in the manufacture of concrete. The amount of water is equally important to produce good concrete. Water helps to flatten cement across the gravel and assist stirring. Research results obtained from testing the water absorption of coarse aggregate (gravel + concrete used) are as follows:

Table 4. Testing results of aggregates water absorption

Examination	**Unit**	**Result**	**SNI Requirements**
Absorption			
a) Fine Aggregate	%	a) 1.52	-
b) Coarse Aggregate	%	b) 15.86	
c) Used Concrete	%		
1) 15 : 85		1) 6.75	
2) 25 : 75		2) 6.83	
3) 35 : 65		3) 6.95	
4) 50 : 50		4) 7.28	
5) 75 : 25		5) 7.65	

In Absorption Testing, it can be seen that the greater the ratio of the used concrete, the greater the resulting absorption. This happens water enters into the pores of the former concrete.

3.5 Fine Modulus and Gradation Limits

Testing sieve analysis is to determine the fine aggregate gradation or distribution of soil particle size *(grainsize* distribution) of a soil sample by using a sieve. According to SNI - 03-2847 - 2002, the fine aggregate was natural sand as a result of the disintegration of the 'natural' rock or sand produced by stone crushers industries and has the largest grain size of 5.0 mm. British Standard (BS) provides gradation requirements for sand. Sand roughness divided into four groups according to the gradation, which is fine sand (zone 4), rather smooth (zone 3), a bit rough (zone 2) and coarse (zone 1) as in table below:

Table 5. Value of FM and gradation limit

No	**Examination**	**Unit**	**Result**	**SNI Requirements**
1	Fine Modulus			
	a) Fine Aggregate		a) 2.33	-
	b) Coarse Aggregate		b) 6.67	

	c) Used Concrete		
	1) 15 : 85		1) 6.49
	2) 25 : 75		2) 6.64
	3) 35 : 65		3) 6.52
	4) 50 : 50		4) 6.43
	5) 75 : 25		5) 6.17
2	Gradation Limit		
	a) Fine Aggregate		a) No.3 -
	b) Coarse Aggregate	mm	b) 40
	c) Used Concrete	mm	
	1) 15 : 85		1) 20
	2) 25 : 75		2) 20
	3) 35 : 65		3) 20
	4) 50 : 50		4) 20
	5) 75 : 25		5) 20

In Testing Sieve analysis can be seen that the greater the comparison on the former concrete then the modulus of fineness obtained getting smaller. According to SNI 03-2461- 1991 has a fineness modulus of fine aggregate or finess modulus (FM), which is in the range between 1.5 s / d 3.8. From the test data obtained fine aggregate are in zone 3 and has a fineness modulus of 2.33. These results are in accordance with SNI 03-2461-1991 which states that the fineness modulus of fine aggregate is in the range of 1.5 s / d 3.8.

4. Conclusions

From the results of physical examination, it can be concluded that are: For the comparison of used concrete starting from the ratio of 50:50 to 75: 25, including one type of lightweight material aggregate because based on SNI 03 - 1969-1990 value of specific gravity> 2,5 gr / cm used for non structural concrete, and if the specific gravity of 2.6 gr / cm to 2.7 g / cm belongs to the normal aggregate used for concrete with a strength of 15 to 40 Mpa. The characteristics of the material in the test require special treatment to be used as a mixture of concrete, such as the need for watering or soaking before use.

Acknowledgement

Thanks to the research and technology ministry of higher education for technology research grants in 2017.

References

[1] Suharwanto. 2005. Perilaku Mekanik Beton Agregat Daur Ulang: Aspek Material-Struktural. Departemen Teknik Sipil Institut Teknologi Bandung. Bandung

[2] Zuhri, I. (2013, September 22). *imamzuhri.blogspot.com.* Dipetik Juni 08, 2016, dari Perilaku Mekanik pada Elemen Struktur Beton Agregat Daur Ulang: http://imamzuhri.blogspot.co.id/2012/09/perilaku-mekanik-pada-elemen-struktur

[3] Harioseto, B., As'ad, S., & Sambowo, K. A. (2013). Pengaruh Komposisi Agregat Daur Ulang Terhadap Susut Pada Beton Normal dan Beton Mutu Tinggi. *e-jurnal Matriks Teknik Sipil, 1*(4), 511-518.

[4] Mulyono Tri, Ir. MT, Teknologi Beton, Andi, Yogyakarta, 2003

[5] SNI 03-6820-2002, *Spesifikasi Agregat Halus untuk Pekerjaan Adukan dan Plesteran dengan Bahan Dasar Semen.*

[6] Rahmadiyanto, Candra; Samekto, Wuryati., *Teknologi Beton*, Penerbit Kanisius, Jakarta, 2001

IOP Conf. Series: Journal of Physics: Conf. Series **970** (2018) 012018 doi:10.1088/1742-6596/970/1/012018

Improve Outcomes Study subjects Chemistry Teaching and Learning Strategies through independent study with the help of computer-based media

Gulmah Sugiharti[1,2*]

[1]Education Technology Doctoral Program, Postgraduate , Universitas Negeri Medan, Medan, Indonesia
[2] Department of Chemistry Education Faculty of Math and Natural Science Universitas Negeri Medan, Medan, Indonesia

*Corresponding author:gulmahsugiharti@yahoo.com

Abstract. This study aims to see the improvement of student learning outcomes by independent learning using computer-based learning media in the course of STBM (Teaching and Learning Strategy) Chemistry. Population in this research all student of class of 2014 which take subject STBM Chemistry as many as 4 class. While the sample is taken by purposive as many as 2 classes, each 32 students, as control class and expriment class. The instrument used is the test of learning outcomes in the form of multiple choice with the number of questions as many as 20 questions that have been declared valid, and reliable. Data analysis techniques used one-sided t test and improved learning outcomes using a normalized gain test. Based on the learning result data, the average of normalized gain values for the experimental class is 0,530 and for the control class is 0,224. The result of the experimental student learning result is 53% and the control class is 22,4%. Hypothesis testing results obtained t count> ttable is 9.02> 1.6723 at the level of significance $\alpha = 0.05$ and db = 58. This means that the acceptance of Ha is the use of computer-based learning media (CAI Computer) can improve student learning outcomes in the course Learning Teaching Strategy (STBM) Chemistry academic year 2017/2018.
Keywords: Self-study, Computer-based Media. Improve learning out

1. Introduction

Teaching and Learning Strategy (STBM) Chemistry is one of the compulsory courses with the weight of 3 credits to be studied by the students of Chemistry FMIPA UNIMED. This course is a prerequisite for students to be able to follow Mikroteaching course (Micro-Teaching) and PPL. A few years experience in STBM Kimia subjects still there are students who failed or did not pass. The number of students who still have not successfully passed, sometimes opened a special class outside the semester.

Students often learn self-taught, the ways they can do since in primary and secondary education, so it becomes a tendency to learn only from lecture notes and diktat when going to take the exam. As a result of such events, it is feared that students can not afford to fulfill the burden of tasks assigned to him after he becomes a lecturer in secondary school later. For that self-learning students need to be improved. Self-learning problems can be overcome in many ways, including with computer-based learning media. Computer-based learning media is called "computer assisted instruction" or CAI. According to Dhani Harda (2016) CAI media is an interactive media that can improve student interaction, can stimulate students' thoughts, feelings, interests, and attention in such a way. In order for the computer can be made interactive learning media and can be used independently need to be designed and developed based on instructional principles. Computer-based learning media or CAI used to improve student learning outcomes in this study is CAI tutorial. The results of the research (Tambunan, Gulmah, 2010) found that 87.83% of students enjoy computer-based independent learning in General Chemistry II courses.

2. Computer Based Media and Independent Learning

The demands of today's educational world have led to enjoyable learning and make it easier for learners to learn. With plan and design that meet the rules and instructional components, the computer can be used as one of the media that can learn individually. Computers will greatly help educators to convey knowladge to learners and conversely with computer learners will be able to learn easily and surprisingly.

According to Shute and Psotka (in Jonassen 1996) computer-based learning makes it easy to describe the contents of the curriculum for someone to learn. The purpose and content of the curriculum can be programmed in such a way by utilizing intelligence machines. In learning computer-aided teaching process is increasingly interactive and helps self-learning.

The forms of presentation of information on the CAI model include: tutorial, drill and practice, simulation and games and others (Arsyad, 2011). Tutorial is a computer-aided teaching program that mimics a tutor's work system. Information of a concept presented on the screen in the form of: text, image or graph. After students have read and considered the information well, a question is asked. If the student's correct answer the computer will present the next concept, if the wrong answer the computer will return to the previous concept or provide remedial. The move to one of the remedial concepts is determined by the mistakes made by the students. The CAI model is good for self-learning because computers can accommodate students who slowly receive lessons, the computer will work according to the desired instruction, the computer can stimulate students to do the exercises because of the availability of graphical visualization, images, colors and music, immediate feedback, can be used without the help of guide teachers, students can learn individually, the control is in the hands of students so that the level of student learning speed adjusted to the level of mastery. the ability to record student activities during program use, so that student development. The CAI tutorial model as proposed by Shute and Psotka in Jonassen (1996) is illustrated as shown below.

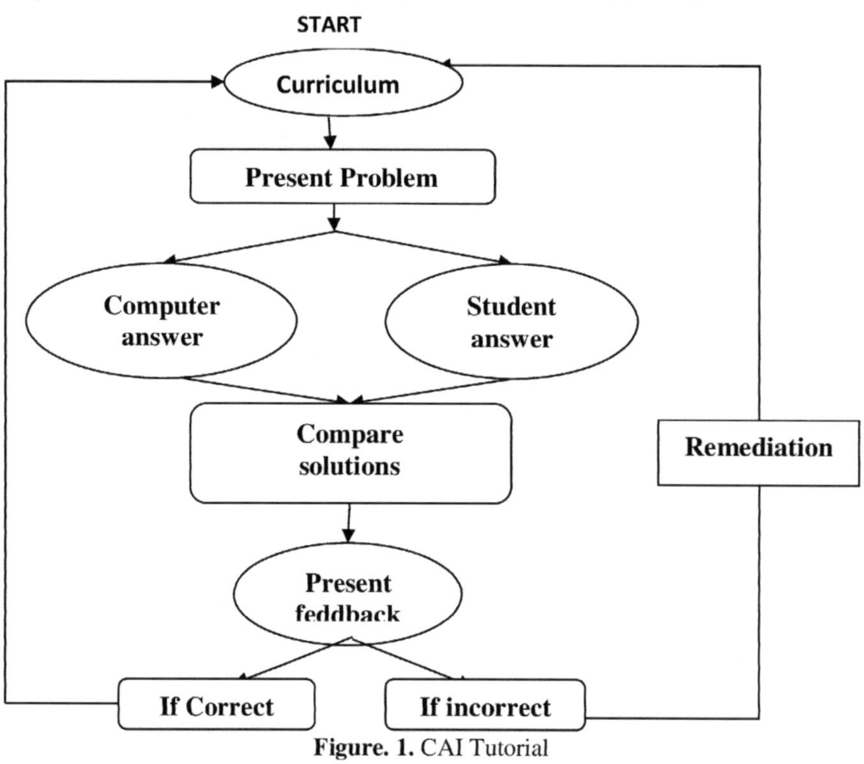

Figure. 1. CAI Tutorial

The procedure includes the material being studied, followed by a problem to be solved as a sub-section of the curriculum. Problem solving was tested by a learner's skill test (in this case a student) displayed through the knowledge and skills instructed at that time. The answer given is compared with the correct answer, then the computer will give feedback. If the answer is correct, a new problem is selected and presented. If the answer is wrong, remediation is given to repeat the learning materials, which are given more simply.

3. Research Methodology

Population in this research is all student of class of 2014 which take subject STBM Chemistry as many as 4 class, and by purposive taken 2 class as sample which each class 32 student. The research design is as follows:

Table 1. Research Plan

Group	Pre test	Treatment	Post Test
Experiments	T_1	X_1	T_2
Control	T_1	X_2	T_2

To see the application of computer-based learning media in improving student learning outcomes in the course Learning Strategy of teaching chemistry, used t test one party is the right side test while

To see the improvement of student learning outcomes taught by using computer-based media can be seen by using gain (gain) with the formula $g = \frac{\text{post test score - pretest score}}{\text{Maximum score -pretest score}}$

By criterion g (normalized gain):

$g < 0.3$ = low, $0.3 \leq g \leq 0{,}7$ = medium, $g > 0.7$ = high

% Improved learning outcomes were calculated through the average gain of the experimental class x 100%.

While the magnitude of the influence of computer-based media can be calculated by the formula;

$$= \frac{gain\ eksperimen - gain\ kontrol}{gain\ eksperimen} x\,100\%$$

$$= \frac{\exp erimentgain - contro\lg ain}{\exp erimentgain} x\,100\%$$

To see Percent (%) increase in learning outcomes used g factor formula (normalized gain gains). The formula used is:

$$g = \frac{Sum\ of\ total\ gain}{Sum\ of\ total\ student}\ x\ 100\%$$

If value g:

<50%: Ineffective media to improve student learning outcomes.

50% - 70: Effective media to improve student learning outcomes.

> 70%: Media is very effective to improve student learning outcomes

4. Research Result and Discussion

4.1 Data of Result Study

Table 2. Table of Average Result Study of Student

Experiment Class						Control Experiment					
PRETES		POSTES		GAIN		PRETES		POSTES		GAIN	
$\overline{X}$	S	$\overline{X}$	S	$\overline{X}$	S	$\overline{X}$	S	$\overline{X}$	S	$\overline{X}$	S
44,00	9,595	75,83	7,08	0,560	0,139	44,00	7,812	57,83	6,524	0,241	0,117

To test the hypothesis, firstly tested the requirement analysis that is the normality and homogeneity test at the significant level ($\alpha = 0{,}05$) and n = 30. While the homogeneity test of variance between group of the comparison sample can be done by equality test of two variance with significant level 0, 05. Normality test results and homogeneity of variance obtained normal distribution data and come from homogenous population.

4.2 Hypothesis Testing Results Student Learning

Hypothesis testing is done by using one party test, that is right side test with $\alpha = 0{,}05$. The calculation results obtained t arithmetic = 9.62 is greater than t table = 1.6723, which means Ho is rejected, meaning the application of computer-based learning media (CAI) can improve student learning outcomes in the course of STBM Chemistry

4.3 Improvement of Student Learning Outcomes

To see the improvement of chemistry learning result of experiment class student and control class seen from mean of normalized gain of experiment class and control class multiplied by 100%. Data on the improvement of students' learning outcomes of experimental class and control class in table 4.

Table 3. Improvement of Student Learning Outcomes Between Experiment Class and Control Class

Sample	**Average Gain**	**% Improvement of Learning Outcomes**
Experiment class	0,530	53,0%
Control Class	0,224	22,4%

Based on the above table it can be seen that the Improved Learning Outcomes of Experiment class students is higher than Control class, can be described in the following graph:

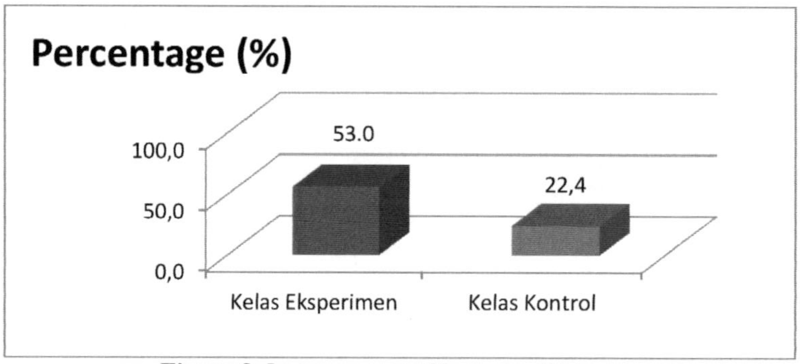

Figure 2. Improvement of Learning Outcome

The results showed that student learning outcomes were taught by using CAI Computer media higher than the student learning outcomes taught without using media. This is because with the application of CAI Computer media, students can more deeply understand the lecture material by using the time available for self study. Things that have been seemingly abstract and difficult to convey through computer-based media (CAI Tutorial) to be interesting and clear.

According to Rusman (2012), with continuous practice, it will be embedded and then will become a habit. Computer-based learning system CAI in the tutorial program is basically the same as the guidance program that aims to provide assistance to students in order to achieve optimal learning outcomes. Thus the learning outcomes can be improved.

From the results of the above data analysis can puladilihat that by using computer-based mediaCAI tutorial provides a new atmosphere to students because of the integration of each sub subject that is taught into the media that makes students easier in understanding the subject matter. Students are actively studying independently using media that gives the learning outcomes of STBM Chemistry course for the better.

Immediate feedback, CAI media developers can design a learning situation that requires students to do something and get positive feedback immediately after students do something (Dhani, 2016). Computers are able to provide information to students about errors and amount of time learning and time to work on problems, computer-assisted learning can also be designed to overcome problems and weaknesses in large group learning. In addition, computer-assisted learning trains students to skillfully select the parts of the desired content of learning. The development of carefully designed learning and psychological attention of students will be useful for students who have difficulty following the traditional method of learning. With Computer Based Learning can provide strengthening to students who are weak in learning because the material or questions can be repeated. Another benefit is computer-assisted learning that allows students to become familiar with and become familiar with computers that are becoming increasingly important in today's modern society.

Based on research that has been done in the study of chemistry education obtained data - data showing that improving learning STBM course through independent learning with the help of computer-based media has been able to improve student learning outcomes.

References

[1] Dhani, H.,. 2016. Pengembangan Media Computer Assisted Instruction (CAI) Untuk Perhitungan Volume Pekerjaan dan Analisis Biaya Bahan Pada Kompetensi Keahlian Teknik Konstruksi Batu Beton. Jurnal Teknologi Pendidikan. 18 (2)

[2] Tambunan, M, Gulmah Sugiharti(2010). *Perbaikan pembelajaran MK Kimia Umum II melalui belajar mandiri dengan bantuan media berbasis komputer pada matakuliah Kimia Umum 2.* Laporan Penelitian PHKI 2010

[3] Jonassen DH (l996). *Handbook of Research for Educational Communications and Tecnology. A project of the association for educational communications and Technology. New York*: Pretice Hall International

[4] Arsyad, A. 2002. *Media Pengajaran*. Jakarta : PT Raja Grafindo Parsada.

[5] Rusman. K, Deni. R, Cepi. 2012. *Pembelajaran Berbasis Teknologi Informasi Dan Komunikasi: Mengembangkan Profesionalitas Guru*. Jakarta: Raja Grafindo Persada.

[6] Surjono, H. (1995). Pengembangan Computer-Assisted Instruction (CAI) Untuk Pelajaran Elektronika. Jurnal Kependidikan. No. 2 (XXV): 95-106

Producer gas production of Indonesian biomass in fixed-bed downdraft gasifier as an alternative fuels for internal combustion engines

J P Simanjuntak[1)], Lisyanto[1)], E Daryanto[1)] and B H Tambunan[1)]

[1,2,3,4)]Mechanical Engineering Department, Universitas Negeri Medan, Medan 20221, North Sumatera, Indonesia

Corresponding author's email : janterps@gmail.com

Abstract-downdraft biomass gasification reactors, coupled with reciprocating internal combustion engines (ICE) are a viable technology for small scale heat and power generation. The direct use of producer gas as fuel subtitution in an ICE could be of great interest since Indonesia has significant land area in different forest types that could be used to produce bioenergy and convert forest materials to bioenergy for use in energy production and the versatility of this engine. This paper will look into the aspect of biomass energie as a contributor to energy mix in Indonesia. This work also contains information gathered from numerous previews study on the downdraft gasifier based on experimental or simulation study on the ability of producer gas as fuels for internal combustion engines aplication. All data will be used to complement the preliminary work on biomass gasification using downdraft to produce producer gas and its application to engines.

1. Introduction

Until today, approximately the most of the world electricity is generated from coal, natural gas, nuclear, and oil. The coal and natural gas play important roles in producing world electricity. However, with the limitation of resource and green house effect, renewable energy source becomes the third popular generation route as energy sources as can be seen in Fig.1. Renewable energy is the energy derived from natural processes that do not involve the consumption of exhaustible resources such as fossil fuels. The exploration for stable energy supplies from various energy alternative resources is important, because most energy resources currently depend on fossil fuels, which are of finite availability. From a technical point of view, the large use of renewable energy sourcess or alternative fuels can also directly donate to improvements in engine performance and polution characteristics. Biomass is one of the alternative energy sourcess and most plentiful organic materials on the earth, which is produced by photosynthesis process in green plant in the presence of sunlight.

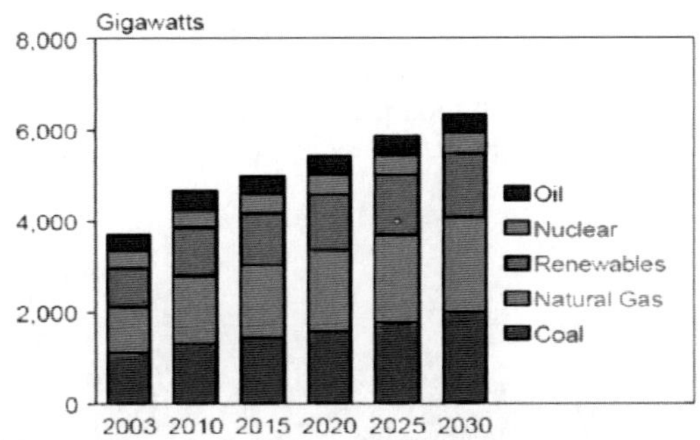

Figure 1. Predicted world electricity generation capacity from 2003 to 2030 [1]

Indonesia has great potential renewable energy resources such as biomass that could be used for further applications. The biomass can be proceed through a gasification process where high temperature and airfuel ratios play important roles. Gasification involve thermo-chemical process for converting a solid fuel into a mixture of combustible gases that can be used for heat and power applications. Downdraft gasifier is one the reactor types that could be used to convert solid fuel. Numerous study on investigation of a downdraft biomass gasifer has been carried out using biomass under various operating conditions. The main advantage of downdraft reactor compare to fluidised bed is the lower tar concentration in the producer gas, which is very important for the durability of the engines and downstream devices. The downdraft gasifier however has some disadvantages suchas limited implementation to small capacities that means difficult to scale-up, lower heating value compared to fluidised bed especially internaly circulating fluidised bed on optimum operation condition (6.79 MJ/m^3) [2]. The largest downdraft gasifiers which exist have the power output in the range from 1.5 MWt to 5 MWt. Similar to world electricity energy, the energy demand in the transportation sector also keeps increasing in the near term where the ratio of biofuel to the total amount of fuel is also expected to increase. Fig.2 shows the fuel consumption needed for world global transportation until year 2040. Gasoline and diesel fuel still the biggest contributor to the world energy of transportation.

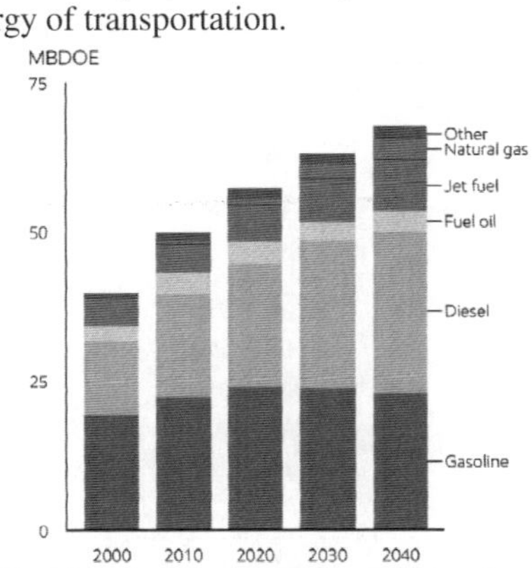

Figure 2. World global transportation deman by fuel [3]

In Indonesia, dynamic changes in the energy sector were happened in 1998 that was called economis crisis periode. Facing such an unfavorable situation, the government of Indonesia prioritizes on energy supply securities by diversification of energy resources. Biomass renewable energy is one of the contributor to energy mix of the other resources such as crude oil, natural gas, coal and renewable energies as shown in Fig. 3. It can be seen that total primary energy supply had increased steadily over the past 19 years. It is estimated to reach about 1270.9 Mboe in 2009 which is more than 200% increase from 1990.

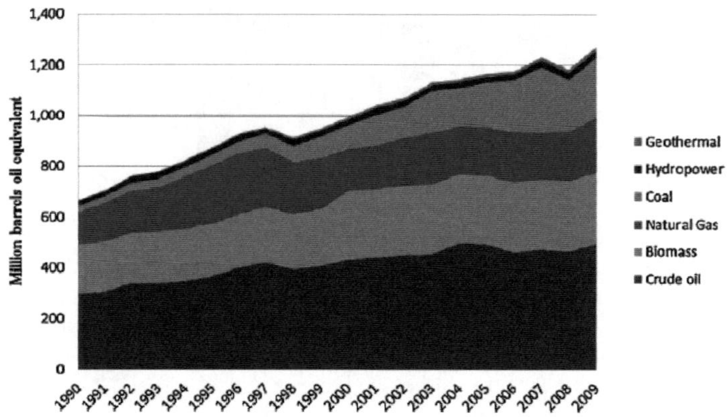

Figure 3. Primary Indonesian energy supply by fuel type in [4]

Nowdays, the energy mix in Indonesia is dominated by petroleum from numerous companies, however almost all the companies reported a drop in the production as shown in Table 1. In this article, it was assumed that gasoline would be substituted by an equal volume of bioenergy due to energy production and the versatility of the engines. In year 2005, Indonesia needs 17.480 billion litre of gasoline to fuel the transportation devices as shown in Table 2.

Table 1. Indonesian primary energy share 2003-2004 [5]

Company	2003	2004	Change (%)
Caltex[a]	506.9	507.0	0.0
Total	81.1	81.8	0.8
CNOOC	94.9	81.5	−14.2
Unocal[a]	53.9	55.7	3.3
Exspan	66.4	54.0	−18.6
Pertamina	43.6	48.4	10.9
Conoco Philips	51.4	44.1	−14.2
Petrochima	40.5	36.6	−9.7
BP	38.8	31.3	−19.4
BumiSiakPusako	32	30.0	−6.2
Vico	32.3	28.8	−10.8
ExxonMobil	25.4	21.2	−16.5
Others	79.6	74.1	−6.9

[a] Changed to Chevron after merger.

Table 2. Fuel consumption for transportation in Indonesia by 2005 (in billion litre) [6]

Fuel type	Transportation	Industry	Household	Electricity	Total
Gasoline	17.480	-	-	-	17.480
Kerosene	-	0.091	11.295	-	11.386

Diesel	12.1322	8.506	-	6.832	27.470
Diesel oil	0.072	0.807	-	0.016	0.895
Fuel oil	0.281	2.[392	-	2.155	4.828
Total	29.965	11.796	11.295	9.003	62.059

2. Potential of biomass as an energy source in Indonesia

Biomass is natural energy source, mostly coming from agriculture crops and residues, forest waste, commodities of plantation, and animal waste. In Indonesia biomass is available in the form of agricultural residues, forestry residues and animal residues. According to [7], estimated total energy that can be obtained from agricultural residues and forestry residues is to be 614 and 141 million GJ/year, respectively. However, utilization of biomass instead of energy sources is suspected to be driving the destruction of forests in Indonesia and create conflict of land use for food or for biomass.

Biomass is the only one of the renewable energy, which can be used to produce three kinds of fuel, liquid, solid and gas fuels. Some of early works on biomass energy conversion in Indonesia are discussed by Refs. The country's total biomass production was around 146.7 million tons per year, equivalent to 470 GJ/y and mostly used by rural areas and small industries to provide energy for cooking, heat and electricity. Biomass energy source in the country can be obtained from palm oil residues, rice residues, rubber wood, palm oil residues, sugar residues, etc. As presented in Table 3. Rice residues have the biggest energy potential about 150 GJ/year, followed by rubber wood, sugar residues, and palm oil residues with energy potential 120 GJ/year, 78 GJ/year, and 67 GJ/year, respectively. The rest are from agricultural wastes and residues from any other agriculture activities such as logging residues, coconut residues, sawn timber residues and plywood and veneer residues with potential energy smaller than 20 GJ/year. Indonesia was known as the largest palm oil producer in the world after overtaking Malaysia in 2006. The country produce 21.6 million tons of palm oil, rise 3.8% from 2009. One of the factors leading to higher production is the expansion of palm oil plantation area by 6.7% to 5.73 million hectare in 2010.

Table 3. Major biomass energy potential as energy sources in Indonesia [8]

Biomass	Main region	Production (Milion/year)	Technical energy potential (GJ/year)	Remarks
Rubber wood	Sumatera, Kalimantan, Java	41 (replanting)	120	Small log<10 cm medium and big logs are used as fire wood in brick and roof tile industry
Loging residues	Sumatera, Kalimantan	45	19	
Sawn timber residues	Sumatera, Kalimantan	13	13	Residues of the factory often used as fire wood by local communities, residue available for free
Plywood and veneer production residues	Kalimantan, Sumatra, Java, Irian Jaya, Maluku	Bagasse: 10 Cane tops: 4 Cane leaf: 9.6	16	Bagasse is generally used in sugar factories (90%), The use of cane leaf and tops need to be investigated
Sugar residues	Java, Sumatra, Sulawesi,	Bagasse: 10 Cane tops: 4	78	Bagasse is generally used in sugar factories

	Kalimantan, Bali/Nusa Tenggara	Cane leaf: 9.6		(90%), The use of cane leaf and tops need to be investigated
Rice residues	Java, Sumatra, Sulawesi, Kalimantan, Bali/Nusa, Tenggara	Husk: 12 Bran: 2.5 Stalk: 2 Straw: 49	150	Stalk and straw are generated at the field and generally burnt, in some areas used for feeding or raw material for paper industry Husks often burnt uncontrolled
Coconut residues	Sumatra, Java, Sulawesi	Husk: 12 Bran: 2.5 Stalk: 2	7	Residues are used generated decentralized and usually left on the plantation field. Largely used as Firewood and for production charcoal
Palm oil residues	Sumatra, new areas: Kalimantan, Sulawesi, Maluku, Nusa Tenggara, Irian Jaya	Empty fruit Bunches: 3.4 Fibers: 3.6 Palm shells: 1.2	67	Palm cells and fiber are common fuel source, EFB are generally incinerated

3. Downdraft fixed-bed gasifier

Fixed bed gasifiers system is the most common technology used for solid biomass conversion. During gasification process, biomass as feedstock are subjected to different reaction zones (drying, pyrolisis, gasification, and combustion). There are many types of fixed bed gasifiers with diverse arrangements. The fixed bed gasifier can also be classified according to the ways the air as the gasifying agent come into the gasifier such as updraft, downdraft, cross draft gasifier. The downdraft application is limited to small capasities. According to [9], the largest downdraft gasifiers which exist have the power output in the range from 1.5 MWt to 5 MWt. The most commercially fixed bed gasifier that used until now is downdraft gasifier, whilst atmospheric downdraft gasifier is an attractive for small-scale application. Table 4 below shows the data of energy content which called heating value from biomass diverse by using fixed bed reactor as the gasifier.

Table 4. Energy content from types of biomass by using fixed bed reactor as the gasifier. [10]

No	Gasifier Type	Biomass Used	Low Heating Value (MJ/Nm^3)
1	Fixed bed downdraft	Wood chips	4,7
2	Downdraft double air stage	Eucaliptus wood	4,6
3	Downdraft gasifier	Sawdust and Sunflower seed	5,6
4	Fixed-bed Twin Fired	Wood chip	5,8
5	Two stage downdraft	Wood chip	6,0
6	Downdraft Gasifier	Olive kernel	4,46
7	Dual fired downdraft	Wood	5,3

8	Downdraft gasifier	Rice husk	5,6
9	Downdraft	Wood chips	5,1
10	Downdraft gasifier	Wood chips	5,52
11	Downdraft gasifier	Wood chips	4,6

4. Syngas for engine application

Internal combustion engines as well as spark-ignition (SI) and compression-ignition (CI) are machines that convert heat produced from fuels combustion into mechanical useful work. The combustion process is one of the most important energy conversion methods where the chemical energy of fuel (producer gas) is directly converted into heat. During the last years, the effect of producer gas, a low-heating value fuels on the engines, (either spark-ignition or compression ignition one) has been intensively studied. There is about 40 % of combustible gases which mainly consists of carbon monoxide (CO), hydrogen (H_2) and some methane (CH_4) is produced since gasification process occured, and the rest are non-combustible and consists mainly of nitrogen (N_2), carbon dioxide (CO_2) and water vapour (H_2O). The are also condensable gas include such as tar, acids and dust which called impurities that could be affected engines operational. Usually, the main problem of gasifier system design is to generate a producer gas with a high proportion of combustible components and a minimum of impurities. Spark ignition engines, normally operated with fuel oil, usually petrol or kerosene, but it can also be run on producer gas alone. Diesel engines can be changed and fuelled by producer gas by lowering the compression ratio and the installation of a spark ignition system. Dual fuel mode is another possibility to run a normal unconverted diesel engine in, whereby the engine draws anything between 0 and 90 % of its power output from producer gas, the remaining diesel oil being necessary for ignition of the combustible gas/air mixture. The advantage of the latter system lies in its flexibility: in case of malfunctioning of the gasifier or lack of biomass fuel, an immediate change to full diesel operation is generally possible. However, not all types of diesel engines can be converted to the above mode of operation. Compression ratios of ante-chamber and turbulence chamber diesel engines are too high for satisfactory dual fuel operation and use of producer gas in those engines leads to knocking caused by too high pressures combined with delayed ignition.

5. Relevant work of syngas aplication on engines

Numerous of the research on the application of low-heating value gaseous fuels in engines has been studied so far throug both modelling and experimental way. Martinez et al. [11] concluded that low energy density of the producer gas/air mixture and the engine's low volumetric efficiency causes power de-rating. However, according to [12] the way to increase the maximum brake torque (MBT) is by simultaneously retarding spark timing and increasing compression ratio. Sahoo et al. [13] reports the iqnition characteristic of producer gas in order to render its long-term use feasible. According to literatures, the engine power de-rating when using producer gas can be as great as 70% [14] . The reason for this decrease is the low calorific value of the producer gas, which rarely exceeds 6–7 MJ/Nm^3 [15]. Another phenomenon related to the use of producer gas in engines, especially spark-ignited ones, is the combustion cyclic variability, which can be quantified by the coefficient of variation (ratio of standard deviation to the mean value) of the maximum combustion pressure or the indicated mean effective pressure (IMEP). The stability of engine operation, as well as the misfire and knocking tendencies, when using producer gas as fuel, have also been studied, using the coefficient of variation of the mean effective pressure [16]. Szwaja et al. [17] studied a SI engine which operates on sewage sludge producer gas mixed with methane in various proportions. The results shows that the coefficient of variation of IMEP was greater than 5%, which is considered unacceptable for power production engines. Similar results are reported by Orbaiz et al, who used four different fuels [18]. By using dry wood as the feedstock [16] also studied the effect of compression ratio and percentage of opened air inlet valve on break power of Honda

Model GX-120-four strokes-spark ignited engine, they used compression ratio of 7.5:1 and 9.3:1 and percentage of opened air inlet valve of 30% and 75%. They found that the brake power of compression ratio at 9.3:1 with 75% of opened air inlet valve was 1,443.6 Watts at 3,800 rpm showed the highest break engine power. Through intake flow of gas modification, the influence of swirl on the performance, emissions and combustion in a constant speed Spark Ignition (SI) engine has been studied experimentally by numerous researcher. A single cylinder diesel engine was also modified to operate as a biogas operated spark ignition engine [19]. They tested engine at 1500 rpm at throttle opening of 25% and 100% at various equivalence ratios. Their tests covered a range of equivalence ratios from rich to lean operating limits and also at an optimum compression ratio of 13:1 with normal and masked intake valve to enhance swirl [19]

6. Conclusion

Through literature study, a downdraft biomass gasifier can be used to generated producer gas instead of fluidised bed that can be applied to fuel an internal combustion engines. Minimum calorific value of producer gas needed is above 3 MJ/m^3 . Considering the relative success of the studies performed in this work, it seems interesting to consider the use of biomass materials to produce syngas for engines. This preliminary work covers potential alternative fuels for automotive engine application for both spark ignition (SI) and compression ignition.

Acknowledgment
The authors would like to acknowledge supports from DIPA, Universitas Negeri Medan, Nomor SP DIPA-042.01.2.400926/2017. Tanggal 07 Desember 2016.

References

[1] Outlook, A.E., *Energy information administration.* Department of Energy, 2010. **92010**(9): p. 1-15.
[2] Simanjuntak, J. and Z. Zainal, *Experimental study and characterization of a two-compartment cylindrical internally circulating fluidized bed gasifier.* Biomass and Bioenergy, 2015. **77**: p. 147-154.
[3] Mobil, E., *The Outlook for Energy: A view to 2040.* Irving, TX: Exxon Mobil, 2012.
[4] Mineral, K.E.S., *Handbook of Energy and economic Statistic of Indonesia.* Center for Data and Information on Energy and Mineral Resources Ministry of Energy and Mineral Resources, p121, 2012.
[5] Embassy, U.J., *Petroleum Report Indonesia, 2005-2006.* Jakarta, Indonesia, 2006.
[6] Suntana, A.S., et al., *Bio-methanol potential in Indonesia: forest biomass as a source of bio-energy that reduces carbon emissions.* Applied Energy, 2009. **86**: p. S215-S221.
[7] Prastowo, B. *Biomass resource in Indonesia: Indonesia's solid biomass energy potential.* in *Proceedings of the Indonesia-German Workshop and Seminar.* 2012: Institute Technology of Bandung.
[8] Hasan, M., T. Mahlia, and H. Nur, *A review on energy scenario and sustainable energy in Indonesia.* Renewable and Sustainable Energy Reviews, 2012. **16**(4): p. 2316-2328.
[9] Beenackers, A., *Biomass gasification in moving beds, a review of European technologies.* Renewable Energy, 1999. **16**(1-4): p. 1180-1186.
[10] Asadullah, M., *Biomass gasification gas cleaning for downstream applications: A comparative critical review.* Renewable and Sustainable Energy Reviews, 2014. **40**: p. 118-132.
[11] Martínez, J.D., et al., *Syngas production in downdraft biomass gasifiers and its application using internal combustion engines.* Renewable Energy, 2012. **38**(1): p. 1-9.

[12] Sridhar, G., et al., *Development of producer gas engines.* Proceedings of the Institution of Mechanical Engineers, Part D: Journal of Automobile Engineering, 2005. **219**(3): p. 423-438.

[13] Sahoo, B.B., N. Sahoo, and U.K. Saha, *Effect of H2:CO ratio in syngas on the performance of a dual fuel diesel engine operation.* Applied Thermal Engineering, 2012. **49**: p. 139-146.

[14] Banapurmath, N. and P. Tewari, *Comparative performance studies of a 4-stroke CI engine operated on dual fuel mode with producer gas and Honge oil and its methyl ester (HOME) with and without carburetor.* Renewable Energy, 2009. **34**(4): p. 1009-1015.

[15] Hernández, J., J. Barba, and G. Aranda, *Combustion characterization of producer gas from biomass gasification.* Global NEST Journal, 2012. **14**(2): p. 125-132.

[16] Orbaiz, P., et al., *A comparative study of a spark ignition engine running on hydrogen, synthesis gas and natural gas.* SAE International Journal of Engines, 2013. **6**(2013-01-0229): p. 23-44.

IOP Conf. Series: Journal of Physics: Conf. Series **970** (2018) 012020 doi:10.1088/1742-6596/970/1/012020

The Effect Of Additional Detergent In Crude Palm Oil In The Process Of Separation Stearin

Vina Rezekyah Hasibuan[1*], Nur aini[1], Febriyanti[1], Salahudin Al Ayubi Pane[1]
[1]Department of Chemistry Education, State University of Medan, Medan, Indonesia

*Corresponding Author: vina.rezekyah@gmail.com

Abstract - This study aims to find out how much stearin is formed from the addition of detergent and to understand the process of separation of crude olein with crude stearin from raw material of crude palm oil (CPO). Using a detergent fractionation system, detergent fractionation is a continuous crystallization of oil with controlled cooling and the separation of fractions by weight or centrifuge after supplementing surfactant.

I. Introduction

Oil palm (elaesis guinensis) is one of the non-oil commodities in the plantation sector that is very supportive of the pace of development. One of the businesses currently in demand by planters and peasants is to grow oil palm to increase income. The processing of oil palm is to obtain the oil from the meat and the core citation of the seeds Palm oil and palm kernel oil are classified into edible oil and can be consumed immediately after the oil has obtained further processing eg after refining and bleaching, because the existing palm oil planted is still in the form of crude palm oil (Crude palm Oil).

Crude Palm Oil (CPO) is an oil derived from crushed palm oil (mesocarp) meat. for the use of crude palm kernel food is generally fractionated in order to separate the liquid oil fraction which contains most of the unsaturated fatty acids of the fraction containing most of the saturated fatty acids. Stearin and olein is a product that is ready to be marketed and has the selling power and its own benefits for the community. From stearin with fractionation and hydrogenation can be made various ingredients for cooking that is shortening, bakery, fat and others. While from stearin with the addition of fatty acid can be made soap, crayons wax surface active agent and stearic acid. From olein with fractionation and hydrogenation can also be made foods that are cooking oil, cooking oil, shortening, margarine, salad oil and biscuit fat.

Commonly, there are 3 CPO fractionation systems, namely dry fractionation, solvent fractionation, detergent fractionation. The detergent fractionation is the batch or continuous cooling crystallization and the separation of the fractions either gravity or centrifugation after the addition of surfactans. The processing of crude palm oil (CPO) in vegetable oil plants aims to improve the function and benefits of palm oil. Where expected oil produced can be consumed by consumers and as raw materials in the factory.

Crude palm oil (CPO) consists of olein fraction and stearin fraction. To get the fraction of olein and stearin is usually done fractionation process by crystallization method. The separation of the crystalline solid material from a solution to precipitate saturated and unsaturated triglycerides. The cause of crystallization is due to the difference of freezing point. Crude stearin freezes at ± 20 ° C in the process of gradual cooling and addition of deterrents to accelerate crystallization.

The use of detergent should not be less than or more than is permitted, because if the use of less than permitted can lead to imperfectly formed crystals so that the separation of the solid and liquid fractions becomes slow, it can also cause the surface of the crystals formed to become coarse and narrow. Whereas the use of more than allowed can result in the amount of material wasted at the time of separation of crude stearin - detergent, can also cause danger to the stearin that is formed later.

2. Method

The working method is done on the separation process of stearin and olein is weigh detergent as much as 0.2 grams then dissolve with 0.8 grams of water, input into beaker glass then add 100 grams of CPO to homogeneous stir. Then place the CPO mixture into the water bath containing the pieces of ice, mix the CPO mixture with a temperature of 10 °C for ± 2 hours until the crystal is formed. After that heat up, to form 2 layers. Then put on for 15 minutes, after centrifying process will form 3 layers, after separated weigh each layer is done density identification test, to be known at which layer there is stearin. After stearin is identified, wash with hot water then input into separating funnel, remove the bottom layer, recalculate the saponification number and do test density test and melting point.

The method of checking the saponification number
Weigh 2 grams of washed sample, feed into 200 ml erlenmeyer, then add 20 ml alcoholic KOH 0.5 N heat into a water bath for 30 minutes. Chill, then add a few drops of PP indicator it will form an old pink color then titration with HCl 0.5 N solution until the pink color is lost. The empty test of the same ordinance will be performed parallel to the full test of the water sample.
Saponizing number formula =
BM KOH x (B-A) r N HCl / Gr Example
Description:
A = Volume HC1 0.5 N at full test/example (ml)
B = Volume HC1 0,5 N in void testing (ml)

Preparation of a standard 0.5 N HCl solution
Pipette 10.36 ml concentrated HCl then dilute into 250 ml measuring flask until the line marks.

Preparation of 0.00N alcoholic KOH solution
Weigh 7.0125 gr KOH dissolved with a small amount of water (as little as possible), then add 95% ethanol to 250 ml measuring flask until line mark, shake until homogeneous, then store for 2 or 3 days.

Method of measuring density
Weigh the 5 ml empty pignometer. then enter the sample, weigh back then the density can be calculated as follows:
ρ = pi-pk / 5 ml

Information:
ρ = Density (gr / ml)
pi = The weight of the pignometer contains the sample (gr)
pk = Empty pignometer weight (gr)
5 ml = volume pignometer (ml)

Method of measuring melting point
Heat the sample in a water bath and measure the temperature at which the stearin begins to melt.

3. Result

Table 1. On the fractionation process

No .	Comparison of amount of CPO (gr) with detergent concentration (%)	Freezing point of crude stearin (°C)
1	100:0,2	10
2	100:0,3	10
3	100:0,4	10
4	100:0,6	10
5	100:0,7	10
6	100:0,8	10

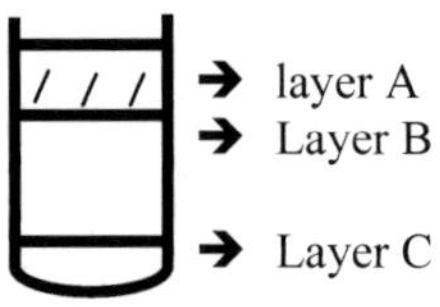

Figure 1. After Centrifuge

Table 2. After Fractionation

No.	CPO (gr)	D + W	After fractination			Loss (gr)
			Layer A (gr)	Layer B (gr)	Layer C (gr)	
1	100	0,2 + 99,8	29,10	68,60	1,12	2,18
2	100	0,3 + 99,7	29,98	67,34	1,27	2,41
3	100	0,4 + 99,6	31,58	65,36	1,23	2,83
4	100	0,6 + 99,4	32,42	64,50	1,41	2,67
5	100	0,7 + 99,3	32,78	63,88	1,63	2,71
6	100	0,8 + 99,2	33,28	63,43	1,67	2,53

Table 3. Result of Stearin Analysis

No.	After washing	Density stearin (gr/ml)	Saponification number		Melting point (°C)
	Pure stearin (gr)		Before washing	After washing	
1	29,02	0,81	201,43	197,08	55
2	29,85	0,815	201,57	197,78	55,5
3	31,21	0,831	201,71	197,92	56
4	32,03	0,85	201,85	198,20	56
5	32,41	0,86	201,99	198,48	56,5
6	32,84	0,863	202,13	198,76	57

in order to solve the main problem then conducted a discussion through the graph based on the relationship between the amount of stearin obtained with detergent concentration.

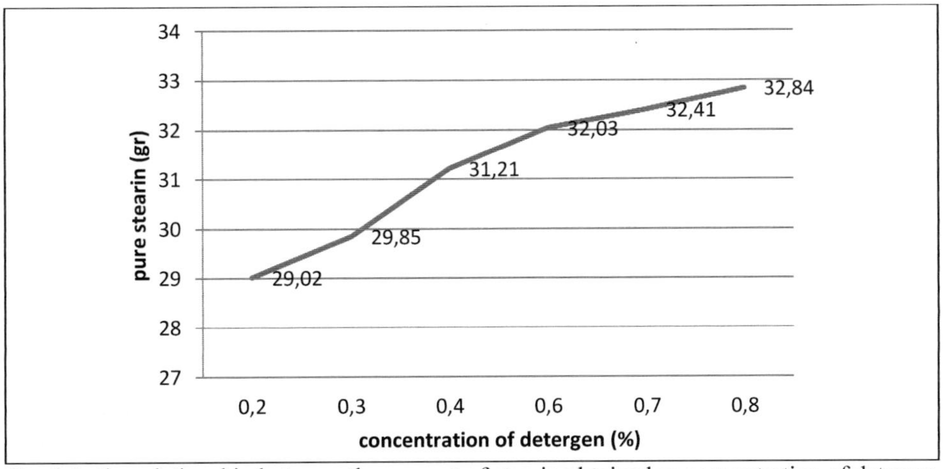

Graph 1. the relationship between the amount of stearin obtained vs concentration of detergen

As for the effect of addition of detergent in CPO in stearin formation process is more and more detergent added into CPO hence stearin gain will be bigger. So that the correlation between detergent concentration with stearin proportion is straight proportional, where in addition of detergent 0,2% obtained stearin as much as 29,02 gr and apada addition of 0,8% detergent obtained stearin as much as 32,84 gr so that detergent usage must be adjusted to that allowed in factory ie 0.6-0.8%.

4. Conclusion

Fractionation by crystallization method using detergent that is NaLS. As for detergent is to accelerate the formation of crystals. in this use is necessary to control the use of detergents, in order to obtain crude stearin in accordance with the desired.

From the results of observations and discussions that have been done then it can be concluded that the effect of adding detergent in CPO on stearin separation process is to add stearin into the CPO can cause a decrease in high freezing point so as to speed up the crystallization process so that if concentration of detergent added into CPO hence gain of stearin will get bigger where in addition of detergent 0,2 obtained stearin as much 29,02 gr and at addition of 0,8 detergent obtained stearin as much as 32,84 gr.

Acknowledgement

The author would like to acknowledge Ir. Mariani Sebayang and Darni Paranita, ST. for supervising the research.

Refferences

[1] Anonim, "Perkebunan dan Pemasaran Minyak Kelapa Sawit Indonesia", PT. International Contact Business System, Inc, Jakarta, 1997.
[2] Anonim, "Pedoman Pabrik Fraksinasi Rafinasi PTPN IV Belawan", buku proses pengolahan di PMN PTPN IV Belawan, 1987.

[3] Fessenden & Fessenden, "Kimia Organik", penerbit Erlangga; Edisi ketiga, jilid 2; university of Montana, page 411-413.
[4] Ketaren S., "Pengantar Teknologi Minyak dan Lemak Pangan" Penerbit Universitas Indonesia (UI Press), Jakarta, 1984.
[5] Lubis Boyke, "Produk Sawit Sebagai Bahan Olah Industri", Buletin Perkebunan. Vol:19 No.1, RISPA-MEDAN, 1998.
[6] Purba Erna, "Penentuan Tekanan Operasi Filter Press FL211D yang Sesuai Untuk Pemisahan Olein & Stearin di Unit Fraksinasi PT. Multimas Nabati Asahan", PTKI-Medan, 2006.
[7] Sari Intan, "Penentuan Keseimbangan NaLS dan Mg SO_4 Terhadap Pembentukan Kristal Crude Stearin di Tangki Kristalizer Pabrik Minyak Nabati PTP Nusantara Belawan", PTKI-Medan, 2001.

IOP Conf. Series: Journal of Physics: Conf. Series **970** (2018) 012021 doi:10.1088/1742-6596/970/1/012021

In Vitro Assay of Ethanolic Heat Reflux Extract of *Nicotiana tabacum* L. var Virginia Against Nosocomial Bacteria Pathogen

Andri Pramono[1], Ahmad Fauzantoro[2], Irma Rizki Hidayati[1], Arina Hygea[1], Oktaviani Sandra Puspita[1], Hikmah Muktamiroh[1], Kristina Simanjuntak[1], Misri Gozan[2,*]

[1] Microbiologi Department, Medical Faculty, Universitas Pembangunan Nasional"Veteran"Jakarta, Indonesia [2] Chemical Engineering Department, Faculty of Engineering, Universitas Indonesia, Depok, Indonesia

*Corresponding Author : mgozan@che.ui.ac.id

Abstract – Tobacco plays an important role in international trade as one of the export commodities. Indonesia is one of the good quality export contributors of tobacco leaves in the world. Nevertheless, tobacco is used only as a raw material for the cigarette industries, and the rise on anti-cigarette regulations prompted the exploration of alternative product from tobacco plants. The content of alkaloids, flavonoids, terpenoids and steroids in tobacco leaves were reported in literatures as antibacterial. Therefore, this study proposed in vitro assay of the ethanolic heat reflux extract (EHRE) of *Nicotiana tabacum* var. Virginia against nosocomial bacteria pathogen ((*Pseudomonas aeruginosa* (ATCC 27853), *Eschericia coli* (ATCC 25922), *Staphylococcus aureus* (ATCC 25923), *Enterococcus faecalis* (ATCC 29212)). Kirby-bauer diffusion method was used for this assay. The concentration of the EHRE for Kirby-bauer assay were 20; 40; 60; 80; and 100%. The presence of clear zones on Kirby-bauer test, against the growth of each nosocomial bacteria pathogen show that tobacco extract has antibacterial effect. Statistical analysis result showed that each extract concentration had significant difference value ($p < 0,05$). This study indicated that the content (alkaloids, flavonoids, terpenoids and steroids) of tobacco leaf extracts (*N. tabacum*) has potential as antibacterial against nosocomial bacteria pathogen. Nevertheless, optimization of tobacco leaf extract to obtain maximum active ingredient still needs to be done. This study is important for further development of the tobacco leaf extract as antibacterial

1 Introduction

Indonesia is one of the world's largest biodiversity nations. There are approximately 30,000 species are spread throughout Indonesia, and about 300 plant species in Indonesia have been used as traditional medicinal materials by the traditional medicine industry.[1] One of the many medicinal plants in Indonesia is tobacco, whose processing is used as raw materials for cigarettes. Tobacco contains 2,500 chemical components. After burning, as many as 1,400 components will be unravel. The unravel component will react and form 4,800 chemical components in cigarette smoke.[2]

The rise of anti-cigarette regulations encouraged necessitate exploration of alternative product of tobacco. Tobacco in Indonesia has an important role, not only in the economic aspects but also in work employment.[3] Besides as raw materials of cigarettes, tobacco was also able as antibacterial,

IOP Conf. Series: Journal of Physics: Conf. Series **970** (2018) 012021 doi:10.1088/1742-6596/970/1/012021

antifungal and bioinsektisida.[4,5,6] Tobacco leaf extract (*Nicotiana tabacum* L.) could inhibited the growth *Bacillus amyloiquefaciens*, *Staphylococcus aureus*, *Pseudomonas aeruginosa* and *Escherichia coli*.[4] But the optimization of antimicrobial compound composition in tobacco extract still require in demand. The antimicrobial inhibition were caused by secondary metabolite chemical compounds like, alkaloids, flavonoids, terpenoids and steroids which are bioactive compounds that act as antibacterial.[7] In addition, tobacco bio-oil pyrolysis extract has been used as a mosquito repellent for humans, since it does not contain DEET (N, N-diethyl-m-toluamide).[7] Excessive exposure of DEET in humans can cause sensory and motor disturbances, and DEET is not found in tobacco extracts.[8] This tobacco potential makes it a potential feedstock for both topical medicines or as a disinfectant in humans.

addition, although the potency of tobacco extract as antibacterial has been tested many times, but the isolation of extract is mostly done by maceration and socletation extraction method, while the isolation by reflux method has not been studied for antibacterial potential.[4,5] The difference of isolation technique lead to different composition of active substance. This led to the importance of an antibacterial test on *Nicotiana tabacum* L. var Virginia tobacco extract by reflux method.

Nosocomial infection will cause the cost of treatment, the stay in the hospital will prolonged or endanger the patient's life. A survey of the prevalence of WHO-managed nosocomial infections, in 55 hospitals in 14 countries divided into 4 regions, namely Europe, Eastern Mediterranean, Southeast Asia and the Western Pacific, showed that approximately 8.7% of hospital patients had nosocomial infections, other surveys suggest about 1.4 million patients worldwide have nosocomial infections. The infection late operation (ILO) could established from nosocomial infections, in the presence of pus, pain, and redness. Pus culture showed various nosocomial bacteria as the cause of infection, both Gram positive and Gram negative. Several researchers have reported the incidence of ILO with 3 types the most bacterial infection, for example in Bangladeshi hospital bacteria identified were *Pseudomonas* sp., *Bacteroides fragilis* and *Escherichia coli*.[14] In RS M. Djamil Padang Indonesia obtained *Klebsiella* sp., *Staphylococcus aureus*, and *Enterobacter aglomerans*, whereas in RS Moewardi Surakarta Indonesia caused by *Enterobacter* sp., *Pseudomonas* sp., and *Proteus* sp.[10] Anaerobic bacteria *Bacteroides fragilis* is also one cause of wound infection.[11]

Traditional medicine has grown globally and its benefits are known to some extent. It expected improve the health care in Indonesia. Here, we tested antibacterial effectivity of tobacco leaf extract (*Nicotiana tabacum* L. var *Virginia*) isolated by reflux method, against some pathogenic nosocomial bacteria with Kirby-bauer diffusion method.

2. Material And Methods

2.1. Time and Place

The time of the research was conducted in March-October 2017. The research was conducted in Microbiology Laboratory of Medical Faculty of National Development University "Veteran" Jakarta. Tobacco extraction was carried out at the Chemical Engineering Department, Faculty of Engineering, and Universitas Indonesia.

2.2. Experiment Design

The type of research used in this study is experimental research with true experiment research design. The research was conducted in laboratory, by giving treatment in various concentration (20%, 40%, 60%, 80% and 100%) tobacco leaf extract (*Nicotiana tabacum* L.var Virginia) and control (aquadest) comparison group. Diffusion method Kirby-Bauer were used to calculate Minimal Inhibition Concentration (MIC). The test group used were *Pseudomonas aeruginosa* (ATCC 27853), *Eschericia coli* (ATCC 25922), *Staphylococcus aureus* (ATCC 25923), and *Enterococcus faecalis* (ATCC 29212).

2.3. Ethanolic Heat Reflux Extract of Nicotiana tabacum L. var Virginia

Tobacco leaves (*Nicotiana tabacum* L.var Virginia) obtained from Ponorogo, East Java. The taxonomy of *Nicotiana tabacum* L.var Virginia was confirmed by the Center for Plant Botanic Gardens, Indonesian Institute of Sciences, Indonesia. To make the simplicia, tobacco leaves were cleaned and washed with water until clean. They were then dried under the sun. In addition for completely drying, the leaves were inserted into an oven with a temperature of 120 ° C for 2 hours. Dried leaf powder is obtained from the grinding of simplicia which is then sieved, to obtain fine powder. The extraction was performed by heat reflux method. A 50 g of the tobacco leaf powder was dissolved into 250 ml of ethyl alcohol. This solution was then placed in a three-neck flask with connected to the Allihn condenser and heated to 80°C. The solution is continuously stirred automatically by a 150 rpm stirrer hotplate for 6 hours. The three neck flask containing tobacco leaf powder and solvent soaked in water bath for heating process to take place evenly. On the side of the Allihn condenser there is a hose connecting the condenser with the cooler (chiller) at 4^{0}C, so condensation process proceed. The vapor formed by heating condensation, and returned to the flask. The mixtures were filtrated under vacuum condition to separate the filtrate from tobacco waste. The filtrate was then evaporated by vacuum evaporator to get concentrated extract.

2.4 Measurement of Antibacterial Activity

Measurement of antibacterial activity was carried out at the laboratory of microbiology FKUPN "Veteran" Jakarta. Kirby-bauer diffusion method was used in this study. The filter paper containing tobacco leaf extract is placed on the surface of a solid medium that previously inoculated by test bacteria on its surface. After 24 hours incubation, the inhibition zone diameter around the disc were measured. Formed around the disc paper was used to measure the resistance strength of extract against the pathogen bacteria.[12]

2.5 Data analysis

One-Way ANOVA test (parametric test) were used to compare the 5 group concentration (0%, 20%, 40%, 60%, 80% and 100%). The data distribution should be normal and the data variance must be the same, to meet the One-Way ANOVA test requirements. Saphiro-Wilk were used to test data normality, while Levene's test were used to test data homogenity. Hypothesis were accepted when a p value <0.05. The Post Hoc analysis test is conducted to determine which groups have significant differences. If the One-Way ANOVA test is not eligible, then the Kruskal-Wallis (non-parametric) alternative test is used. [13]

3. Result And Discusssion

3.1. The tobacco extract showed inhibitory effect on all concentration against all 4 bacteria

Based on the results, the effect of tobacco leaf extract (*Nicotiana tabacum* L.) against all four bacterial nosocomial pathogen (*P.aeruginosa* ATCC 27853, *Eschericia coli* ATCC 25922, *Staphylococcus aureus* ATCC 25923, *Enterococcus faecalis* ATCC 29212). The inhibitory power shown by the formation of clear zones around disk paper in experiments on 0%, 20%, 40%, 60%, 80% and 100% extract concentration. Any data in this method was obtained by four-replications (Figure 1.).

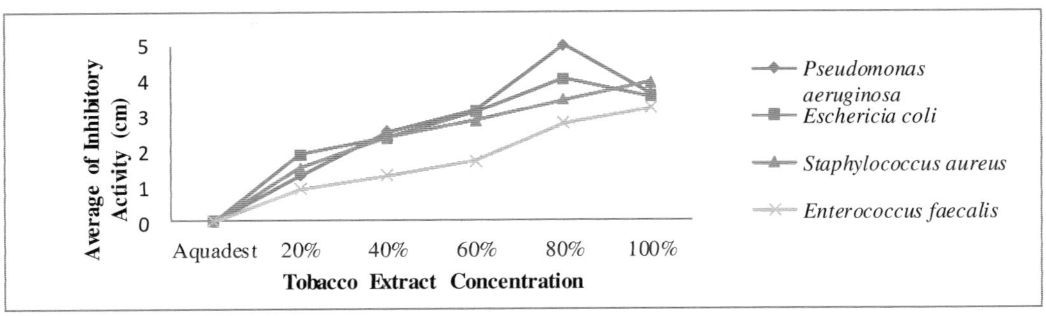

Figure 1. Average Inhibitory Activity of Ethanolic Heat Reflux Extract of *Nicotiana tabacum* L. var Virginia Against Nosocomial Bacteria Pathogen

This study showed that tobacco leaf extract (*Nicotiana tabacum* L.) has an antibacterial effect on *P.aeruginosa* ATCC 27853, *E.coli* ATCC 25922, *Staphylococcus aureus* (ATCC 25923) and *Enterococcus faecalis* (ATCC 29212). All result against 4 bacteria, were included in the weak category.

3.2. Statistical analysis showed significance difference on all extract concentration with some exception

Table 1. Statistical analysis MIC result of different concentration tobacco extract

Tobacco concentration	extract	*P. aeruginosa* Significance	*E. coli* Significance	*S. aureus* Significance	*E. faecalis* Significance
20%	40%	0,000	0,001	0,021	**0,083**
	60%	0,000	0,000	0,021	0,020
	80%	0,000	0,000	0,021	0,021
	100%	0,000	0,000	0,021	0,021
40%	20%	0,000	0,001	0,021	**0,083**
	60%	0,000	0,000	0,021	0,020
	80%	0,000	0,000	0,021	0,021
	100%	0,000	0,000	0,021	0,021
60%	20%	0,000	0,000	0,021	0,020
	40%	0,000	0,000	0,021	0,020
	80%	0,000	0,000	0,021	0,020
	100%	0,007	0,004	0,021	0,020
80%	20%	0,000	0,000	0,021	0,021
	40%	0,000	0,000	0,021	0,021
	60%	0,000	0,000	0,021	0,020
	100%	0,000	0,001	**0,248**	0,021
100%	20%	0,000	0,000	0,021	0,021
	40%	0,000	0,000	0,021	0,021
	60%	0,007	0,004	0,021	0,020
	80%	0,000	0,001	**0,248**	0,021

One-Way ANOVA test of *P.aeruginosa* ATCC 27853 and *E.coli* ATCC 25922, showed that significance value of p <0,05 meaning tobacco leaf extract (*Nicotiana tabacum* L.) had inhibitory effect as antibacterial. Post-Hoc Test Results (LSD) showed all groups in *P. aeruginosae* and *E. coli*, had significant inhibitory differences (p<0,05) on both bacterium (Table 1.). The extract concentration of 80% yielded the largest mean inhibitory zone diameter in *P.aeruginosa* ATCC 27853 and *E.coli* ATCC 25922. However, when tested with 100% extract concentration, the mean value of the resulting inhibit zone decreased. From these results, it can be seen that the concentration of tobacco leaf extract is most effective in inhibiting the test bacteria is the concentration of 80% extract. At 80% extract concentration, high active ingredient content can be absorbed into paper and media disc to optimally and effectively, so that the drag zone diameter is large (Table 1). This phenomenon happened due to the high concentrated and high consistency of tobacco extract. The diffusion process of extracts on paper discs and media would be less effective, when the consistency of tobacco extract almost solid[14]

Our result is in line with previous research, that concentration of 20%, 40%, 60%, 80% and 100% tobacco extract against *Staphylococcus aureus* and *Escherichia coli* bacteria, has an antibacterial power.[4,7,11] The highest inhibitory were showed in 100% concentration on both *S. aureus* and *E. faecalis* (Gram positive), which was different from *P. aeruginosae* and *E.coli* (Gram negative), which was on 80% concentration (Figure 1.). Probably, this is due to the complexity of bacterial plasma membrane of Gram positive bacteria group, which was more moderate that Gram negative bacteria group.

Post Hoc test on *Staphylococcus aureus* (ATCC 25923) and *Enterococcus faecalis* (ATCC 29212) with Kruskal-Wallis showed that there were difference of various tobacco leaf extract concentration, except concentration 80% with 100% against *S. aureus*, and 20% with 40% against *E. faecalis*.

3.3. Phytochemical result of ethanolic heat reflux tobacco extract

The activity of inhibitory power of tobacco leaf extract (*Nicotiana tabacum* L.) to pathogen bacteria were caused by phytochemical content of tobacco extract, that acts as antibacterial, such as alkaloids, flavonoids, terpenoids and steroids.[5] The phytochemistry result of our ethanolic heat reflux extract showed that the extract composed of alkaloid, saponin, tanin, fenolik, flavonoid, tritefenoid and glycoside (Table 2.).

Table 2. Qualitative phytochemistry result test of EHRE of *Nicotiana tabacum* L. var Virginia

Type of testing	Result	Type of testing	Result
Alkaloid	+	Flavonoid	+
Saponin	+	Tritefenoid	+
Tanin	+	Steroid	-
Fenolik	+	Glycoside.	+

The phytochemical content in the tobacco leaf extract gives antibacterial effect to each of the test bacteria with different mechanism. *Aeruginosa* and *E.coli* bacteria were included in Gram-negative bacteria, whereas *S. aureus* and *E. faecalis* were included Gram-positive bacteria. Gram-negative bacteria have a layered cell wall structure, consisting of lipopolysaccharides, lipoproteins and peptidoglycan. Alkaloid content can interfere with peptidoglycan components in bacterial cell walls. The disruption of peptidoglycan formation makes the lining of the cell wall unintegrated resulting in cell death.[15] In addition, the aromatic compounds contained in the alkaloids cause the formation of bonds with bacterial DNA so that bacterial DNA synthesis is disrupted.[42] The flavonoid content of the tobacco leaf extract may bind to proteins cell membranes that cause cell membrane damage. The destruction of the cell membrane results in disruption of the integrity of the cell membrane of the bacteria resulting in the disturbance of bacterial cell growth and bacterial death.[11] The content of steroids and terpenoids, have an antibacterial ability by inhibiting the action of enzymes involved in energy production and changing the composition of constituents cells due to the accumulation of lipophilic components, thus disrupting the formation of bacterial cell walls. The tobacco extract also contained phenol. High levels of phenol can cause protein denaturation and cause bacterial cell lysis.

This study showed that tobacco leaf extract (*Nicotiana tabacum* L.) isolated by this reflux method still resulted in small MIC effect, compared to previous studies that used tobacco leaf extract (*Nicotiana tabacum* L.) [5,16] The extract improvement still need to be done to recover high amount of antibacterial active compound, to improve growth inhibition on nosocomial bacteria.

References

[1] Depkes RI. 2007. *Keputusan Menteri Kesehatan Republik Indonesia tentang Kebijakan Obat Tradisional Nasional*. Jakarta.

[2] Tirtosastro, S., Murdiyati, A.S. *Kandungan Kimia Tembakau dan Rokok.* 2010. Buletin Tanaman Tembakau, Serat dan Minyak Industri. 2010 April; 2(1): 33-43.

[3] Nurifah, E & Darmadji. 2003. Prospek Pasar Tembakau Jawa Timur. *Jurnal Widya Agrika*, Vol. 1, no. 2, hlm. 77-89, Agustus 2003.

[4] Sharma, Y., Dua, D., Nagar, A., Srivastava, N.S. 2016. *Antibacterial Activity, Phytochemical Screening And Antioxidant Activity Of Stem Of Nicotiana Tabacum*. IJPSR. 2016 Mar 01; 7(3): 1156-1167.

[5] Puspita, P.E. 2011. *Aktivitas Antibakteri Ekstrak Tembakau Temanggung Varietas Genjah Kemloko.* Fakultas Teknologi Pertanian, Institut Pertanian Bogor.

[6] Susilowati, E.Y. 2006. *Identifikasi Nikotin Dari Daun Tembakau (Nicotiana Tabacum) Kering Dan Uji Efektivitas Ekstrak Daun Tembakau Sebagai Insektisida Penggerek Batang Padi*

(Scirpophaga innotata). Tugas Akhir II Untuk Memperoleh Gelar Sarjana Sains, Fakultas Matematika Dan Ilmu Pengetahuan Alam, Universitas Negeri Semarang.

[7] Jufri M, Irmayanti E, Gozan M. 2016. Formulation of Tobacco Based Mosquito Repellent to Avoid Dengue Fever. International Journal of PharmTechResearch. 9 (7) :140-145.

[8] Gozan M, Yasman, Wulan PDK , Ellen Dawitri. 2014. Tobacco leaves pyrolysis for repellent active compound product. International Journal of Applied Engineering Research. 9, 21 : 9739-9749.

[9] Ahmed MI. Prevalence of Nosocomial Wound Infection Among Postoperative Patients and Antibiotics Patterns at Teaching Hospital in Sudan. *North American Journal of Medical Sciences*. 2012;4(1):29-34

[10] Raihana N, 2011, Profil Kultur dan Uji Sensitifitas Bakteri Aerob dari Infeksi Luka Operasi Laparotomi di Bangsal Bedah RSUP Dr. M. Djamil Padang, Fakultas Farmasi Universitas Andalas Padang

[11] Ayub M, Rizwan H, Siddique S, Maryam U. 2015. Isolation of Pathogens Causing Sepsis, Pus and Infected Wounds from Critical Care Unit: A Retrospective Study. Annals of Clin and Lab Res.3(4): 50.

[12] Hudzicki, J. 2009. *Kirby-Bauer Disk Diffusion Susceptibility Test Protocol.*

[13] Dahlan, M.S. 2011. *Statistik untuk Kedokteran dan Kesehatan.* Salemba Medika, Jakarta.

[14] Putra, IA, , Erly, Masri, M 2015, Uji Efek Antibakteri Ekstrak Etanol Kulit Batang Salam {Syzigium polyanthum (Wight) Walp} terhadap Staphylococcus aureus dan Escherichia coli secara Invitro*', Jurnal Kesehatan Andalas*, Vol. 4, no. 2, hlm. 497-501, 2015.

[15] Permatasari, GAAA, Besung, INK, Mahatmi, K 2013, 'Daya Hambat Perasan Daun Sirsak Terhadap Pertumbuhan Bakteri Escherichia coli', *Indonesia Medicus Veterinus*, Vol. 2, no.2, hlm. 162-169, Januari 2013.

[16] Putri, A.A., Rasyid, R., Rahmatini. 2014. *Perbedaan Sensitivitas Kuman Pseudomonas Aeruginosa Penyebab Infeksi Nosokomial Terhadap Beberapa Antibiotika Generik Dan Paten*. Jurnal Kesehatan Andalas. 2014; 3(3): 327-331.

IOP Conf. Series: Journal of Physics: Conf. Series **970** (2018) 012022 doi:10.1088/1742-6596/970/1/012022

Review Of Implementation In Bunut Shoes Assistance Program In Order Of Micro, Small And Medium Enterprises Economic In Asahan Regency

M. Saleh Malawat[1] **and M. Umar Maya Putra**[2*]

[1]Management Department, University Of Asahan, Kisaran, Indonesia
[2]Management Department, University Of Azhar, Medan, Indonesia

[*]Email: UMAR_YAZLI@YAHOO.COM

Abstract. This paper studies the implementation of business opportunities that can improve the revenue of Bunut Shoes Micro, Small and Medium Enterprises. Probit model with E Views 6 program was used to see how far the opportunity of variable efforts to improve the revenue such as education, training, capital assistance, technological procurement of them. The data used was the primary data by conducting a survey using questionnaires to members of them with the observation period from 2013 to 2015. The results showed that all variables of implementation did not have a business opportunity correlation to the increase in revenue and Asahan District Governments are asked to create a creative breakthrough in order to achieve optimal business revenue and cooperate with other private institutions related to increase the business income.

1. Introduction

The development of Micro, Small and Medium Enterprises (MSMEs). at the time of the 1997 in the monetary crisis era, has no resilience of big business especially in relation to exports. MSMEs feels difficult to get a better level due to the resilience of the stable economic conditions. From Figure 1, it can be seen from 2007 to 2011 although there is an increase of export for it but the export amount is less result than big business so it is necessary to do a special study MSMEs to be able to compete in face of the development of the era.

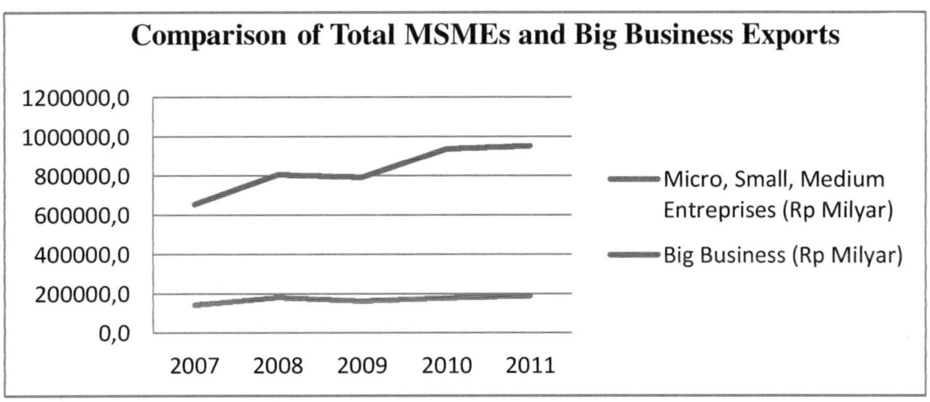

Figure 1. Graphic Of Comparison of Total MSME and Big Business Exports

IOP Conf. Series: Journal of Physics: Conf. Series **970** (2018) 012022 doi:10.1088/1742-6596/970/1/012022

Source: [1]

Problems commonly faced by most MSMEs in Indonesia is the level of productivity in producing output, the quality of human capital that affects the business income of MSMEs less than the maximum. [2] In economic theory, productivity is a measure of output or worker productivity is a maximum ability of a worker to produce output. In reality, the worker does not necessarily take full advantage of his ability and how far the worker utilizes his ability to be measured by the productivity figures. From the phenomenon, then the main problem here is how the size of the productivity of educational investment results that will provide increased quality of human capital.

Increased human capital is expected to improve performance better. [3] explains that in order to respond to change, many companies claim that human capital has a competitive advantage that can improve company performance. This causes the company to understand that human capital will increase workers' satisfaction and improve performance.

Nevertheless, there is an assumption that human capital has a positive effect for the performance of the firm. It is necessary to test a particular company's feasibility in order to be clearly understood. Human capital which is the main point in improving the quality of MSMEs is required an entrepreneurship education support to be able to increase innovation so as to be able to improve the performance of the company. The MSMEs needed the capability in combining the income, capital, selling price, the amount of manpower from the community who can solve all problems to reach the market share globally.

In conducting global marketing a company is required not only to market to consumers a form of product but also there is a service element in the offering of a product even today the wider community more need the main service role than support services. Services play an active role in contributing to the economy of a country [4].

In relation to developing target markets to be able to make each MSMEs got more focused in creating the solution to the problems such as revenue, entrepreneurship education, training, capital assistance and the help of the ideal technology in developing productivity are expected to improve the strategy to solve all problems incorporated in it. From the various studies, researchers were interested in making these five variables become the analyze of the Bunut Shoes MSMEs earnings probability.

Indonesia which has a lot of population that can be used as capital in improving economic development. The large number of residents can certainly be seen in various factors both negative and positive. [5] firstly to pay attention to the positive effects of population development that can be concluded that the increase of population and the provision of education to them before becoming a workforce, enables a society to acquire not only skilled labor, but also skilled, educated, and An educated entrepreneur.

Associated with skilled workers and entrepreneurs, two major theories back to the choice of entrepreneurial choice have emerged. [6] First, the expected utility view claims that individuals choose entrepreneurial choices when they expect a higher yield of workload relative to wages. Second, the view of non-monetary benefits argues that individuals choose entrepreneurship even when returns are lower than for non-pecuniary benefits.

Entrepreneurship which may be a missing link in contemporary growth models corroborates with recent empirical studies that have found an empirical regularity in the form of a positive relationship between various steps of entrepreneurial activity, most typically start-up costs and indicators of economic growth [7].

Previous research has shown that, although small firms are generally less involved in research and development than large firms so they tend to use their innovative inputs more efficiently than large companies. To improve regional development, a change that is usually mentioned in the invention and innovation is required [8].

Invention and innovation in the business world is basically closely related to the strategy of industrial companies in controlling the state of the market. Large industrial firms run the invention and innovation is generally always superior to win the sympathy of consumers in the market [9].

Academic research has highlighted the role of entrepreneurship and venture capital in stimulating innovation. Hundreds, if not thousands, of papers have examined the relationship between firm size and innovation. Much of this work has related measures of innovative discoveries-for example, Research and Development (R & D) expenditures, patents, or invention to firm size [10].

The number and strength of the ties that a Venture Capital (VC) has with its peers and its portfolio investments. This is captured by a specific structural network attribute, that is, the number of connections the VC has (also accounting for the fact that a connection with the same partner may occur repeatedly). Ceteris paribus, the more information the VC has available to use, filter, or recombine, the higher the likelihood that it can extract the value from this information [11].

In the middle stage of economic development, government investment is still needed to increase economic growth in order to take off, but at this stage the role of private investment is getting bigger [12].

2. Method

The research location is located at Jl Jend. Sudirman Bunut Village Asahan Regency. Population taken by 20 Bunut Shoes MSMEs who started business from 2013 to 2015. The sample selection was conducted by the whole sampling where the researcher assigned.

Data source in this research is done by using primary data and secondary data. Primary data sourced from the respondents of the study obtained based on the results of questionnaires and interviews in the form of income of MSMEs, entrepreneurship education, Capital, Training and Technology assistance while secondary data obtained on the general description of Bunut Shoes MSMEs and Library research Supporting theories relevant to the research.

Operational Definition Variable This research is:

- MSMEs revenue is income or yield derived from sources of income received by MSMEs measured by unit 0 for income down, 1 for income up from 2013 to 2015.
- Entrepreneurship education through learning activities to prepare SMEs to take a specific course of action applied in the place of business and assist participants in improving achievements as measured by the number of frequencies from 2013 to 2015.
- Capital is business development assistance obtained from Asahan Regency Government as measured by the number of frequencies from 2013 to 2015.
- Training is to create shoe making skills with the size of the frequency of activities through business mentoring and seminars as measured by the number of frequencies from 2013 to 2015.
- Technological Assistance is the number of governments providing effective technology assistance to an increase in operating revenues as measured by the number of frequencies from 2013 to 2015.

Descriptive Analysis Method That is the analysis used to analyze data by way and Model Analysis Probita model that dependent variable studied follow the normal cumulative distribution. Because it is based on the normal cumulative distribution function, this model is also referred to as the normit model). [13] The probit regression model is an analytical method used to describe the relationship between predictor variables and response variables of more than 1 category. The probit regression is the abbreviation of Probability Unit based on the normal cumulative normal probability distribution function known as Normit model as the Normal Probability Unit. The shape of the probit model is:

$$\textbf{SMEs Bunut Shoes Revenue} = \beta_{j0} + \beta_{j1}\textbf{Education} + \beta_{j2}\textbf{Capital} + \beta_{j3}\textbf{Training} + \beta_{j4}\textbf{Technology}_4 + e$$

SMEs Bunut Shoes Revenue = 0 (Revenue Down), 1 (Revenue Up)

β_{j0} = Constanta

$\beta_{j1}, \beta_{j2}, \beta_{j3}$ = Coeffisien Of Education, Capital, Training and Technology that was counted with the duration of the activity

Hypothesis Test on this research used Goodness Of Fit statistic test by Likelihood Ratio (LR) statistics test, Partial Test Z (Z-Test) and Mc Fadden Coefficient Determination.

IOP Conf. Series: Journal of Physics: Conf. Series **970** (2018) 012022 doi:10.1088/1742-6596/970/1/012022

3. Result of Research

Asahan District is an area that has natural resources that can be used to improve the welfare of the community. Bunut Shoes is one result of handicraft industry that is directly made by the son / daughter of the area. Since 1987, the result of bamboo shoe crafts (Bunut shoes) is a product that has good prospects and quality guaranteed. Made from leather, attractive models and good durability make Bunut shoes have devoted buyers / consumers.

Gallery bunut shoes located at a strategic location on the road Sumatra that is very advantageous because it is easily affordable and competitive prices with other famous brand shoes. In 2010, when the government provided subsidies / assistance in the form of small loans, especially for MSMEs managed by people who have small capital, the name of Bunut shoes re-emerged in the market of shoe industry in the region so that in the year 2013-2015, The Gallery Bunut shoes that had begun to dim, looked excited again, but after doing research in the field found that there is a probit model equation:

MSMEs Revenue = -4.071142 + 1.142652**Education** - 0.544562**Capital** + 0.042448**Training** + 0.825838**Technology**

Table 1. Probit Model Analysis Results From E Views 6

Dependent Variable: REVENUE
Method: ML - Binary Probit (Quadratic hill climbing)
Date: 06/12/17 Time: 19:29
Sample: 1 20

Included observations: 20
Convergence achieved after 5 iterations
Covariance matrix computed using second derivatives

Variable	Coefficient	Std. Error	z-Statistic	Prob.
EDUCATION	1.142652	0.755615	1.512214	0.1305
CAPITAL	-0.544562	0.673235	-0.808874	0.4186
TRAINING	0.042448	0.605712	0.070080	0.9441
TECHNOLOGY	0.825838	0.812434	1.016498	0.3094
	-4.071142	3.787699	-1.074833	0.2824

McFadden R-squared	0.551477	Mean dependent var	0.850000
S.D. dependent var	0.366348	S.E. of regression	0.291885
Akaike info criterion	0.879190	Sum squared resid	1.277951
Schwarz criterion	1.128123	Log likelihood	-3.791899
Hannan-Quinn criter.	0.927784	Restr. log likelihood	-8.454182
LR statistic	9.324565	Avg. log likelihood	-0.189595
Prob(LR statistic)	0.053480		

Obs with Dep=0	3	Total obs	20
Obs with Dep=1	17		

From the model estimation results in the table obtained LR Statistic value of 9.324565 with a probability level of 0.053480. This shows that the independent variables are jointly entrepreneurship education, capital, training and giving technology affect the dependent variable that is MSMEs revenue shoes.

From the results of Partial Test Z conducted to see the effect of independent variables on the dependent variable in one by one, where:

1. Entrepreneurship education has a value of z statistic: 1.512214 with regression coefficient of 1.512214 with probability: 0.1305 which has insignificant value at the level of $\alpha = 5\%$.
2. Capital has the value of z statistic: -0.808874 with regression coefficient of -0.544562 with probability: 0.4186 is not significant at the level of $\alpha = 5\%$
3. The training has a z statistic value: 0.070080 with regression coefficient 0.042448 with probability: 0.9441 which has no significant effect on the level of $\alpha = 5\%$.
4. Technology has a value of z statistic: 1.016498 with regression coefficient of 0.825838 with probability 0.3094 which has no significant effect on the level of $\alpha = 5\%$.

Coefficient of Mc Fadden Determination (R^2) is used in the rate of 0.551477 or 55,1477% to see how many contributions of independent variable that is entrepreneurship education, Capital, Training, Technology can simultaneously explain dependent variable that is MSMEs Revenue while 45 percent explained other variable.

Based on the calculation and explanation, it can be seen that all the variables do not have a significant result to increase the income of MSMEs, it is necessary to further discuss each variable that is:

1. Entrepreneurship Education

The average of the shoe seller is still in the new category in this business and still penetrated in the shoe business due to the threat of getting a job. The inequality of Asahan regency government in providing entrepreneurship education 7 times and at least 2 times causing this variable insignificant affect the revenue of MSMEs. As the investigation [2], productivity is a measure of output or worker productivity is a maximum ability of a worker to produce output. It means that the duration of workers' education should be added more frequency.

2. Capital

Just as entrepreneurship education also experience problems with the division of all MSMEs. Problems that often occur MSMEs less able to determine the business plan and fear if the payment of debt required a warranty payment by asserting assert as collateral. Frequency of capital given by the government 7 times and on average get help 2 times. The recommendation from [3],[4] about the company needs to understand the human capital satisfaction and performance improvement about product knowledge in business plan should be needed by the MSMEs.

3. Training

Business training provided by the government of Asahan Regency such as marketing, shoe making, still felt less than all people. Only a few get 6 times the frequency of training and the majority gets 1 training so that this training pattern answers the role of residual factor of technical progress in economic growth of Asahan Regency and need to identify process of training needed by MSMEs as a whole. As the information [5] that population and the provision of education to them before becoming a workforce, [6], [7] need the entrepreneurial choices and [8], [9], [10] should make the invention and innovation

4. Technology

Appropriate technology that is a help to MSMEs in the form of productivity support tool is expected to give impact to the development of MSMEs, but the aid 6 times done still not give significant results for the development of MSMEs. In this case technology as [11] investigation will make the value from this information and [12] will increase economic growth in Asahan.

IOP Conf. Series: Journal of Physics: Conf. Series **970** (2018) 012022 doi:10.1088/1742-6596/970/1/012022

4. Conclusion

Business development seen from the income of Bunut Shoes MSMEs need to be reviewed from 4 variable which are interrelated among entrepreneurship education, capital, training and technology which is the assistance of government of Asahan Regency. The effectiveness of this aid, can be seen from the given frequency of the probit model used as a quantitative measure.

From this probit model we get an input that all the variables of entrepreneurship education, capital, training and technology are all not running significantly to increase business income, this is because the need for a new improvement started from 2013 as the impact of less attention of this Bunut shoes business from year to year. Asahan District Governments asked to create a creative breakthrough in order to achieve optimal business revenue for MSMEs and cooperate with other parties related to evenness of business income.

In the process of increasing MSMEs income for the future, it needs an in-depth study by taking other variables that are interconnected well and other models that can be developed to produce a better research conducted by other researchers.

References

[1] http://www.depkop.go.id/berita-informasi/data-informasi/data-umkm/. (2013). *http://www.depkop.go.id*. Retrieved 09 30, 2017, from http://www.depkop.go.id.

[2] Bahri, Z. (2011). Analisis Produktifitas Investasi Sumber Daya Manusia Melalui Sarana Pendidikan dan Pelatihan di Kota Jambi. *Jurnal Paradigma Ekonomika Fakultas Ekonomi Universitas Jambi Vol 3 No. 1* , 30-41.

[3] Marimuthu, M., Arokiasamy, L., & Ismail, M. (2009). Human Capital Development And Its Impact On Firm Performance: Evidence From Development Economics. *The International Journal Of Social Research* , 1-8.

[4] Bhayangkari, S. K. (2012). Penerapan Standarisasi Dan Adaptasi Sebagai Upaya Pengembangan Strategi Produk Pada Pemasaran Internasional. *Mankeu, Vol. 1, No. 3* , 245-258.

[5] Sukirno, S. (2011). *Ekonomi Pembangunan Proses, Masalah dan Dasar Kebijakan.* Jakarta: Prenada Media Group.

[6] Tamvada, J. P. (2010). Entrepreneurship and Welfare. *Small Bus Econ* , 65–79.

[7] Braunerhjelm, P., Zoltan, Audretsch, D. B., & Carlsson, B. (2010). The missing link: knowledge diffusion and entrepreneurship in endogenous growth. *Small Bus Econ* , 105–125.

[8] Stam, E., & Wennberg, K. (2009). The roles of R&D in new firm growth. *Small Bus Econ* , 77–89.

[9] Teguh, M. (2010). *Ekonomi Industri.* Jakarta: RajaGrafindo Persada.

[10] Lerner, J. (2010). The future of public efforts to boost entrepreneurship and venture capital. *Small Bus Econ* , 255–264.

[11] Alexy, O. T., Block, J. H., Sandner, P., & Ter Wal, A. L. (2012). Social capital of venture capitalists and start-up funding. *Small Bus Econ* , 835–851.

[12] Lailan, S., & Rahayu, S. E. (2011). Analisis Pengaruh Investasi Pemerintah Dan Swasta Terhadap Penciptaan Kesempatan Kerja Di Sumatera Utara. *Jurnal Manajemen & Bisnis*, 1-11

[13] Kurniasari, L., Sumarminingsih, E., & Solimun. (2010). Permodelan Regresi Logistik Dan Regresi Probit Pada Peubah Respon Multinomial. *Jurnal Universitas Brawijaya* , 309-312.

IOP Conf. Series: Journal of Physics: Conf. Series **970** (2018) 012023 doi:10.1088/1742-6596/970/1/012023

The Antifungal Inhibitory Concentration Effectiveness Test From Ethanol Seed Arabica Coffee (*Coffea arabica*) Extract Against The Growth Of Candida albicans Patient Isolate With In Vitro Method

Adam Satria Rakatama[1], Andri Pramono[2,*], Retno Yulianti[3]

1.Medical Undergraduate Program, Universitas Pembangunan Veteran Jakarta,
2.Microbiologi Department, Medical Faculty, Universitas Pembangunan Veteran Jakarta,
3. Pathology Department, Medical Faculty, Universitas Pembangunan Veteran Jakarta
*Corresponding email: andri.pramesti@upnvj.ac.id

Abstract-*Candida albicans* are the most frequent cause of Vulvovaginalis Candidiasis infection. Its treatment using antifungal drugs, are oftenly caused side effects. The reduction of *C.albicans* growth and the reduction of antifungal drugs side effect, were our main purposed. Our study objective is determine the effectiveness of inhibitory power of arabica coffee seed ethanol extract on the growth of *C.albicans* patient isolates. The type of this research is experimental research. Kirby-bauer method with the Saboraud Dextrose Agar (SDA) media was used in this experiment. Inhibitory zone was observed around the disc, to determine the inhibitory power. The results showed that the inhibitory zone was formed on arabica coffee seed ethanol extract on 10%, 20%, 40%, and 80% concentration. Kruskal-Wallis test results ($p<0,05$) showed that there was a significant difference in mean between the concentration groups tested against the treatment group. The inhibitory zone was formed because of biochemical compound in arabica coffee seed such as caffeine, phenol, alkaloids, flavonoids, and saponins. Inhibitory zone in *C.albicans* patient isolates were smaller compared with *C.albicans* ATCC 90028 as gold standard. This showed that the virulence of *C.albicans* from patients isolates were higher. We concluded that arabica coffee seed ethanol extract could inhibiting the growth of *C.albicans* patient isolates. Optimization of coffee seed ethanol extract to obtain maximum active ingredients still needs to be done. This knowledge is expected to be used for the beginning manufacturer antifungal drug from natural product.

1. Introduction

Infectious diseases caused by fungus are known as mycosis and the highest incidence of mycosis is Candidiasis. Candidiasis is a fungal infection caused by *Candida* species. There are more than 200 species of *Candida* and the most pathogenic *Candida* is *Candida albicans.*[1]

One of the most common types of Candidiasis is vulvovaginalis Candidiasis. Vulvovaginalis Candidiasis is the most common disease in the Dermatology and Venereology clinic RSUD Dr. Soetomo Surabaya with 65.4%.[2] In 2013, as many as 13 patients (56.5%) of 23 pregnant women at Haji Adam Malik Hospital Medan found cases of vulvovaginalis Candidiasis.[3]

Various natural ingredients that exist in the community turned out to have an useful alternative as a herbal treatment against *C.albicans*.[4] One of the natural ingredients known to have an ability to inhibit the growth of *C.albicans* is coffee beans. Coffee beans could be selected because they were widely used in the community, their side effects are relatively lower than synthetic drugs, and they easy to obtain.[5]

Coffee beans are divided into two types, arabica coffee beans (*Coffea arabica*) and robusta coffee beans (*Coffea canephora*).[6] Arabica beans are the most popular coffee beans in the world. Arabica coffee beans are many and easy to find in the wider community, so the price is cheaper than robusta coffee.[7]

Recently, studies have shown that coffee beans have therapeutic effects such as anti-inflammatory, antifungal, and antibacterial.[8] Prior research with boiling technique, with aquadest solvent, and using dilution method found that arabica coffee beans have an inhibitory effect to *C.albicans* stronger than robusta coffee beans with 1 ml of arabica coffee beans giving $20,3x10^4$ CFU / mL inhibition.[9]

In this study, we collected *C.albicans* isolates that were found in Candidiasis patient because they had higher pathogenity and virulence compared with pure strains such as ATCC.[10] Here, we examine the antifungal inhibitory concentration effectiveness test from ethanol seed arabica coffee (*Coffea arabica*) extract against the growth of *C.albicans* isolate from patient with in vitro method.

2. Material And Method

2.1 Design, Place and Time

The type of research used in this study is an experimental study with the research design used is posttest only control group design. This research was conducted at Microbiology Laboratory Faculty of Medicine, Universitas Pembangunan Nasional "Veteran" Jakarta. Preparation of ethanol extract of arabica coffee beans (*Coffea arabica*) was done at Balai Penelitian Tanaman Rempah dan Obat (BALITTRO), Bogor. The study was conducted in July 2017.

2.2 Number and Sampling Method

C.albicans isolate samples were taken from patient from Dermatology Venereology and Obstetric Ginecologic Clinic at RSPAD Gatot Subroto Jakarta, which was diagnosed clinically or suspected Vulvovaginalis Candidiasis. Sampling was conducted during July 2017 and was conducted for one month.

The sample of each group is calculated using Federer's formula with n is the number of samples, and t is the number of groups. The formula used is as follows: $(n-1)(t-1) \geq 15$
Based on the formula, each sample is repeated as many as four times.

2.3 Materials and Tools

The tools used include digital screw term, hand scoon, test tube, bunsen, disc paper size 6 mm, 100 pieces, Bekker glass, petri dish, syringe, centril oce, measuring cup, autoclave, aluminum foil. The ingredients used were ethanol extract of arabica coffee beans, Saboraud Dextrose Agar (SDA), CHROMAgar Candida, C.albicans suspension from patient isolate and *C.albicans* ATCC 90028 culture, 100 ml sterile distilled water, 10% sterile NaCl 0.9%, ketoconazole tablet 200 mg, $BaCl2$ 1% 0.5 ml, $H2SO4$ 1% 99.5 ml.

2.4 Research procedure

Sampling of *C.albicans* isolates was performed after the patient read the informed consent and filled the respondent sheet. Sampling is carried out by the medical officer at the

Presidential Hospital RSPAD Gatot Soebroto Jakarta. After the specimen was taken and inserted into a tube containing NaCl, the researcher gave the tube identity and brought it to the Microbiology Laboratory of the Medical Faculty of the National Development University "Veteran" Jakarta.

The isolate sample was grown into CHROMAgar Candida medium, then incubated with 37^0C for 48 hours to determine colonization. The Candida species can be determined using CHROMAgar Candida according to the manufacturer's instructions that produce the media. After that, the green-colored isolates on CHROMAgar media are performed differentiation test with an incubation on SDA medium with temperature of 45^0C for 48 hours. This process aims to distinguish between *C.albicans* and *C.dubliniensis* because they have a color similarity in the CHROMAgar Candida media.

Once confirmed the growing colonies were C. albicans, the colony was taken and suspended C.albicans isolates patients with a standard of 0.5 McFarland. The suspension is applied to the SDA media, then give the disc paper that has been immersed in each test group. Then the media was incubated in 37^0C for 1x24 hours. After incubation is complete, the Kirby-bauer method were performed by observed the inhibitory zone around the disc paper.

2.5 Data analisis

Inhibitory zone data were analyzed by using Kruskal-Wallis test. If the results of the analysis show significant differences between the control group and the treatment group, then the Post Hoc

Mann-Whitney test is followed. Data processing and data analysis is done using SPSS for Windows version 17.0.

3. Results And Discussion

In this study, seven patients who have been diagnosed with vulvovaginalis Candidiasis diagnosed and are willing to take samples by health workers at Presidential Hospital RSPAD Gatot Soebroto Jakarta. In the differentiation test results using CHROMAgar Candida media, we obtained six samples from patients that contain colonies of *C.albicans* and one sample of patients there are no colonies of *C.albicans*.

The measurement results of this experiment have different sizes, the smallest inhibitory starts at a concentration of 10% and the greatest inhibitory at 80% concentration (Table 1.) The inhibitory zone formed showed the effectiveness of inhibition of ethanol extract of Arabica coffee beans (*Coffea arabica*) on the growth of *C.albicans* isolates. The result showed that the negative control group has no inhibitory zone. The negative control group had no effect in inhibiting the growth of *C.albicans*. While the ethanol extract group of Arabica coffee bean (*Coffea arabica*) and the positive control group showed an inhibitory zone, where the barrier started to form on ethanol extract of Arabica coffee bean (*Coffea arabica*) with concentration of 10% and the diameter increased according to addition of concentration of extract tested (Table 1.). The experimental results show that there is a zone of resistance to the growth of *C.albicans* patient isolates. Zone is formed clear but still visible presence of faint patches around the paper disc which is a fungus metabolite.

Table 1. Average Measurement Results On C.albicans Patient Isolate and ATCC 90028

Repetition	Inhibitory zone on Candida patient isolate (mm)						Inhibitory zone on Candida ATCC 90028 (mm)					
	10%	20%	40%	80%	Control (+)	Control (-)	10%	20%	40%	80%	Control (+)	Control (-)
1	0.61	1.46	2.74	4.55	12.74	0	1.56	2.41	3.69	5.5	14.04	0
2	0.70	1.49	2.44	4.70	13.42	0	1.65	2.44	3.39	5.65	14.37	0
3	0.63	1.62	2.20	4.43	13.54	0	1.58	2.57	3.15	6.02	14.49	0
4	0.69	1.33	2.58	4.46	13.34	0	1.64	2.43	3.53	5.41	14.29	0
Σ	0.65	1.47	2.49	4.53	13.26	0	1.56	2.41	3.69	5.5	14.04	0

The average inhibitory zone produced from ethanol extract of Arabica coffee bean (*Coffea arabica*) 80% concentration is 4.53 mm (Table 1). This concentration provides the greatest inhibitory effect of the antifungal effectiveness of ethanol extract of Arabica coffee beans (*Coffea arabica*) against the *C.albicans* isolates tested. At concentrations of 40%, 20% and 10% respectively give an inhibit zone average of 2.49 mm, 1.47 mm and 0.65 mm.

The result with *C.albicans* ATCC 90028 showed broader inhibition compare with than the *C.albicans* patient isolates. This occurs due to the virulence from patient isolate which is more higher than *C.albicans* ATCC 90028.

Ethanol extract of Arabica coffee bean (*Coffea arabica*) 80% concentration has the greatest inhibition power compared with other concentrations, while extract with the smallest inhibitory power is the concentration group of 10%. In dilution of ethanol extract of Arabica coffee beans (*Coffea arabica*) from concentration of 100% to 80%, 40%, 20% and 10%, there

is reduction of dissolved active substances in each of these concentrations, therefore the inhibitory effectiveness is lower with the smaller concentration of the extract tested. This is evident from the large difference in inhibitory diameter of growth of *C.albicans* isolate, where with the lower concentration of extract tested, the inhibitory power diameter is smaller.

According to Davis and Stout (1971) the power of antifungal inhibition is divided into 4 categories, namely weak, medium, strong and very strong.[11] Based on these criteria, the result of diameter zone inhibition of ethanol extract of Arabica coffee bean (*Coffea arabica*) on *C.albicans* isolate at concentrations of 10%, 20%, 40%, and 80% was classified into the weak inhibitory category (≤ 5mm). This is in contrast to the results of ethanol extract test of Arabica coffee bean (*Coffea arabica*) with *C.albicans* ATCC 90028 which has moderate inhibition category (5-10 mm) at 80% concentration.

This may be due to the *C.albicans* isolate having higher virulence than the ATCC 90028 strain, so the inhibitory power generated at the isolate growth is lower than that of the pure strain.

The higher the concentration of extract tested, the more active substances contained. Conversely, the lower the concentration of the extract, the less active substance, so that the inhibitory power effectiveness will be reduced.[12]

The concentration of the extract affects the speed of the diffusion of a nutritious substance. The greater the concentration of extract, the faster the diffusion occurs, so the greater the inhibitory power and the greater the diameter of the inhibitory zone formed.

Kruskal-Wallis test results have a significance value of 0.000. Because of the significance value <0.05, from the results of this test it can be concluded that there are significant differences in the results between the concentration groups tested against the growth of *C.albicans* patient isolates.

The data analysis was followed by a Mann-Whitney post-hoc test to identify significant differences on different treatment groups. Mann-Whitney test statistics showed that each group of ethanol extract of Arabica coffee bean (*Coffea arabica*) had a significant difference. It can be seen from the significance value less than 0,05 (Table 2.)

Table 2. Mann-Whitney Post-hoc Test

Treatment Group		Sig.
Positive Control	10%	0.000
	20%	0.000
	40%	0.000
	80%	0.000
10% Concentration	20%	0.000
	40%	0.000
	80%	0.000
20% Concentration	40%	0.000
	80%	0.000
40% Concentration	80%	0.000

The anti-fungal effect of Arabica coffee bean extract (*Coffea arabica*) is caused by chemical compounds in Arabica beans (*Coffea arabica*). The chemical compounds include classes of compounds such as caffeine, phenol, alkaloids, flavonoids, and saponins (Table 3.).

Alkaloid compounds affect the isolates of *C.albicans* by inhibiting the biosynthesis of fungal nucleic acids, prevented the fungus to develop and eventually die. In addition to alkaloids, phenols also play a role in inhibiting the growth of *C.albicans* isolates by denaturing the protein bonds in the cell membrane and making the cell membrane lysate and allowing the phenol to penetrate into the cell nucleus and inhibit the *C.albicans* development.[13] This is because *C.albicans* is a very sensitive on phenol compounds.[14]

Flavonoids works by disrupting the formation of pseudohifa during the course of pathogenesis and saponins work by affecting the permeability of molds and yeasts of *C.albicans*. In addition, flavonoids are also able to inhibit the action of enzymes that play a role in the synthesis of *C.albicans* cell walls such as manan synthase, chitin synthase and glucan synthase. If the enzyme is inhibited activity then the formation of fungal cell wall will be disrupted and the formation of pseudohifa will be inhibited.[15]

Table 3. Phytochemical Test Results Arabica Beans

Phytochemical Test	Arabica coffee bean extract
Alkaloids	+
Saponins	+
Tannin	+
Phenolic	+
Flavonoids	+
Triterpenoids	+
Steroids	-
Glycoside	+

Saponins are antifungal by binding to the ergosterol present in the cell wall of the fungus, so the permeability of fungal cells will be disrupted and may inhibit the growth and even death of the fungal cells.[15] Then, there is a tannin that works by interfering with the permeability and causing the shrinking of the cell wall of the fungus that can lead to the death of the fungal cell.[15]

The solvent used in this study is ethanol solvent because ethanol is a safe and good universal solvent for the extraction process of all secondary metabolites, so it can dissolve all active ingredients contained in a natural material such as coffee beans .[16]

4. Conclusion

The ethanol extract of Arabica coffee beans (*Coffea arabica*) has an inhibitory effectiveness on the growth of isolates of *Candida albicans* by disc diffusion method. Then, there was a significant difference in each concentration of ethanol extract of Arabica coffee bean (*Coffea arabica*) in inhibiting the growth of *Candida albicans* isolate. Then, the largest concentration (80%) of ethanol extract Arabica beans (*Coffea arabica*) is the most effective concentration in inhibiting the growth of *C.albicans* isolates.

References

[1] Brooks, GF, Butel, JS, Morse, SA 2012, *Mikrobiologi kedokteran Jawetz, Melnick & Adelberg*, EGC, Jakarta

[2] Tasik, NL, Kapantow, GM, Kandou, RT 2016, Profil kandidiasis vulvovaginalis di poliklinik kulit dan kelamin RSUP Prof. Dr. R. D. Kandou Manado periode Januari-Desember 2013. *Jurnal e-Clinic*, vol.4, no.1, pp.207–214.

[3] Ryanti, D 2013, *Prevalensi Kandidiasis vulvovaginalis dan Vaginosis bakterial pada ibu hamil di poli ibu hamil RSUP Haji Adam Malik Medan periode Agustus – Oktober 2013*. Universitas Sumatera Utara, Medan.

[4] Agnita, P, Waluyo, J, Wahyuni, D 2014, *Perbedaan daya hambat ekstrak dan rebusan daun jarak pagar (Jatropha curcas L.) terhadap pertumbuhan Candida albicans (Robin) Berkhout*. Jember University, Jember.

[5] Rachmawati, W, Winarsih, S, Nurdiana 2013, *Pengaruh ekstrak etanol daun kopi robusta (coffe robusta lindl) terhadap pertumbuhan Candida albicans secara invitro*, Brawijaya University, Malang.

[6] Virgita, A 2012, *Peran Kopi dalam Mencegah Karies*, Sriwijaya University, Palembang

[7] Rahardjo, P 2012, *Kopi*, Penebar Swadaya, Jakarta

[8] Antonio, A, Farah A, dos Santos, KRN, Maia, LC 2011, The potential anticariogenic effect of coffee. *Science Against micriobial pathogens: communicating current research and technological advances*.

[9] Yowanda, I 2015, *Perbandingan daya hambat biji kopi robusta (Coffea canephora) dan arabika (Coffea arabica) terhadap pertumbuhan Candida albicans,* Universitas Syiah Kuala, Banda Aceh.

[10] Lestari, PE 2010, Peran faktor virulensi pada patogenesis infeksi Candida albicans.*Stomatognatic*, vol.7, no.2, pp.113–117.

[11] Davis, WW, Stout, TR 1971, Disc plate method of microbiological antibiotic assay, *Applied Microbiology*, vol.22, no.4, pp.659-665.

[12] Asti, S 2015, *Pengaruh ekstrak biji kopi robusta (Coffea robusta) terhadap aktivitas fagositosis sel monosit*, Jember University, Jember.

[13] Wulandari, A 2012, *Uji daya efektivitas antifungi ekstrak biji tanjung (Mimusops elengi Linn.) terhadap pertumbuhan Candida albicans secara in vitro dengan metode difusi*, Universitas Pembangunan Nasional "Veteran" Jakarta, Jakarta.

[14] Puspitasari, D 2015, *Uji efektivitas ekstrak biji avokad (Persea americana Mill.) terhadap pertumbuhan Candida albicans secara in vitro dengan metode difusi cakram*, Universitas Pembangunan Nasional "Veteran" Jakarta, Jakarta.

[15] Gauniyal P dan Teotia U 2014, Antimicrobal activity and phytochemical analysis of ethanolic extracts of twelve medicinal plants against oral micro organisms, *International Journal of Pharmaceutical and Medicinal Research*, vol.2, no.1, pp.21-27.

[16] Raut, JS, Chauhan NM, Shinde, RB, Karuppayil, SM 2013, Inhibition of planktonic and biofilm growth of *Candida albicans* reveals novel antifungal activity of caffeine. *Journal of Medicinal Plans Research*, vol.7, no.13, pp.777–782.

IOP Conf. Series: Journal of Physics: Conf. Series **970** (2018) 012024 doi:10.1088/1742-6596/970/1/012024

A Study of Work Based Learning For Construction Building Workers

Syafiatun Siregar[1]

Student Doctoral Program Universitas Negeri Padang

Email: syafiatunsiregar@gmail.com

Abstract. Work-based learning (WBL) is designed to improve the competence of participants. This study aims to apply the WBL and to develop attitudes, knowledge, skills, behaviors, and habits, which in turn can improve the competence of construction workers in the field to be sampled. This research was conducted on building construction workers in Medan City with 30 research subjects. The results showed that the evaluation of learning increased in phase I obtained the difference of the average score of 20.9 (the meeting I) and 25.50 (meeting II). The final result shows that the level of activity and competence increased significantly after WBL

1. Introduction

Concrete work on building construction is the science that underlies the development of technological infrastructure especially building infrastructure. The concrete work itself has various types that must be controlled by construction workers such as foundation work, wall mounting work, column work, beams and plates and finishing. The finishing work consists of draining work, ceramic installation, printing techniques/relief and others. The development of technology is so rapidly building construction work that requires the competence and expertise of construction workers to get a strong, sturdy and aesthetic value. Therefore, construction workers must have adequate competence/skills before entering their workforce.

Data of informal workers in Medan city are workers with education level up to high school until 2016 almost reach 34,32%. The high number of informal workers is an obstacle for the government in providing formal employment. Likewise, for informal workers will have difficulty entering the formal workforce due to its very high competitiveness. Therefore, the work most accessible to informal workers without having to show skills as a construction worker. The obstacle faced is that construction workers have low knowledge in entering the world of building construction work. Low knowledge will then be related to the low wages that will be accepted by them. To overcome these problems, the need for a lesson that can improve the competence of construction workers.
Competence of finishing expertise from concrete work among others ceramic installation, wall plastering, and building relief/molding technique is a job that has high wage value according to the level of difficulty of its work. If finishing work is provided with the education and training of construction workers, it will have an impact on increasing competence, selling power in the labor market, competition in the world of work, increasing wages and ultimately impacting welfare. This competency is highly valuable if workers in the field are competent to the concrete finishing work.

[1] Departement Of Civil Engineering Universitas Negeri Medan

IOP Conf. Series: Journal of Physics: Conf. Series **970** (2018) 012024 doi:10.1088/1742-6596/970/1/012024

Competence is the ability that must be owned by a person to perform a task or work based on knowledge, skills and work attitude in accordance with the required performance (Djatnika, 2005). With the competence of a person then the concerned will be able to do a job, organize it so that work can be done, make adjustments when something happens different from the original plan and use the ability it has to solve problems and perform tasks with different conditions. Based on the competencies described above where concrete work that is part of the finishing work is a job that requires special skill competence. Finishing work requires a lot of cost in the process. Therefore, to avoid the mistakes of construction finishing work is to improve the quality of production and improve the competence of workers in their work.

The purpose of this research are:
a. To implement work-based learning on construction workers,
b. To streamline a good learning process, supervised so that the result is technically correct (process/stages)
c. To improve the competence/expertise of construction workers on concrete finishing work

2. Literature Review

Research conducted by Eraut (Linehan, 2008: 14) on educational research, especially at tertiary level, explains the gap between knowledge needed in the workplace and the knowledge and skills generated through formal education. Eraut classifies the knowledge of professional and vocational education programs as follows: a) theoretical knowledge, b) methodological knowledge, c) practical and technical skills, d) general skills, and e) general knowledge of the work concerned. Eraut states that of the many knowledge provided only a few that fit the workplace.

Various theories put forward to define the notion of work-based learning (WBL). Ball, I (2010: 4) describes the definition of the WBL as follows:
a. Vocational workers/learners undertake to learn to improve professional qualifications in their work
b. Workers/students undertake a general study using their type of work to learn and work to assess.
c. Workers conduct in-house programs which are then assessed by the owner of the company
d. Workers undertake planned programs to integrate learning and workplace practices

Boud and Garrick (1999, in Ball I, 2010: 4) states that the WBL is a workplace lesson not only related to direct workplace competence but is an investment in assessing the technical capabilities of the technical workers and the specific knowledge and capabilities in wherever they work. Work-based learning lies in the context of a paradigm shift from 'industrial society' to 'knowledge society' (Rohlinetal, 1998 in Linehan, 2008: 20). Little, (2006) declared the WBL as all forms of learning through the workplace, whether the work experience or work shadowing within a certain time. Another definition states that WBL is all learning that occurs as a result of workplace activity.
Work-based learning as a learning approach plays a role in promoting professional development and learning. WBL is used as a terminology in various countries for programs at school/college to gain experience from the world of work (WBL Guide, 2002). Also used for teenagers to be ready for the transition from school to work to learn the reality of the work/world and be ready to make the right choice in work (Paris & Mason, 1995). Another notion, states: "Work-based learning is any training that relates directly to the requirements of the job on offer in your organization" (Glass, Higgin, & McGregor, 2002).

Work-based learning is a learning strategy that connects learners directly with work in the workplace. This is in line with Bragg (1995):
Work-based learning means instructional programs that deliberately use the workplace as a site for student learning. Work-based learning programs are formal, structured, and strategically organized by instructional staff, employers, and sometimes other groups to link learning in the workplace to students' college-based learning experiences. Work-based learning programs have formal instructional

plans that directly relate students' worksite learning activities to their career goals. These experiences are usually but not always college-credit generating.

Work-based learning is useful in improving learners' motivation for achieving learning outcomes, improving personal and social competence (life skills), improving their understanding of the world of work. Work-based learning is an increasingly important part of the higher education curriculum. Work-based learning can spur the development of student personality and professionalism so they are ready to enter the workforce. Work-based learning tends to be a direct training of learners in the world of work. This learning is more structured by using a work-tailored curriculum.

Bailey & Merrit, 1997, states that recent research results conclude that the use of Work-Based Learning Approach (WBL) in education has a positive influence on achievement, motivation and continuing education WBL. Research and evaluation show the correlation between outputs and graduate outcomes with learning structures provided by schools and industry as workplace experience.

The development of learning in the implementation of vocational education should continue to be done by the managers so that the quality of graduates in accordance with the demands of the labor market. Challenges in the work world with higher work competencies as technology advances and workplace dynamics require vocational education institutions to anticipate and cope with the changes that occur by utilizing various capabilities. Various signals about the quality of education are the background to the problem of vocational education with the WBL approach. WBL is a learning approach that utilizes the workplace to structure workplace experiences that contribute to the social, academic, and career development of learners and be a supplement to learning activities. With the WBL, learners develop attitudes, knowledge, skills, enlightenment, behaviors, habits, and associations from the experiences of both places and allow for learning related to real-life activities (Lynch & Harnish, 1998).

3. Methodology

This research will be conducted on existing construction workers in Medan City with 30 respondents. The place of research was conducted at Civil Engineering Workshop of Building Engineering Department. The implementation of WBL learning does not have to be the same stage, but this stage is adapted to the type of learning, organizational readiness, facilities and infrastructure facilities, funds, human resources available. The steps that must be carried out in a WBL are a preliminary stage, implementation stage, observation stage and re-planning phase. The detailed steps are as follows: 1) identifying needs; 2) formulate objectives; 3) design instructional analysis; 4) developing methods; 5) determine the evaluation pattern; 6) implement the program and; 7) measure learning outcomes.
The text of your paper should be formatted as follows:

4. Result and Conclusion

The results of the implementation of this research activity are described in accordance with the stages in the form of learning cycles conducted in the process of learning in the classroom. Each cycle consists of four stages: planning, execution, observation/reflection, and replanning. The planning stage is done by planning the process before the learning such as identification of needs, formulate goals, design instructional analysis, preparation of learning tools, group determination, materials, and others. The implementation stage is done with three meetings for plastering wall, ceramic installation and molding technique which is done by introduction stage, core stage, and closing stage. The preliminary stage is to pre-test the participants. The core stage is by giving the material by the structure and explaining the learning that will be done. Observation stage where the instructor records all activities during the learning process while at the end of the learning evaluation.

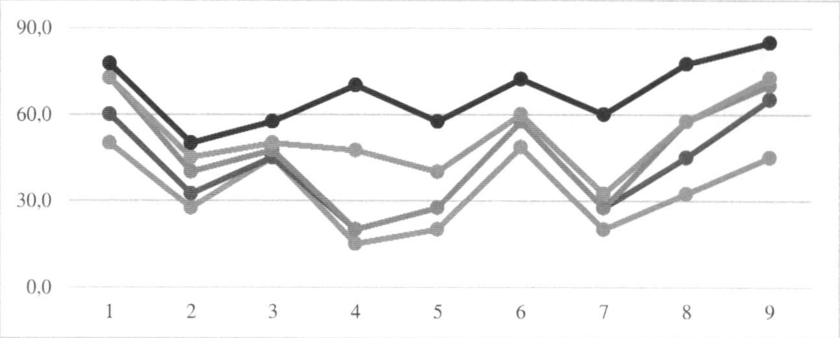

Figure 1. The percentage of the level activity of participants

Graph 1 above is the percentage of active participants in following the WBL. The x-axis shows the measured aspects of the WBL participants consisting of 9 aspects: 1) Noting explanation of instructor, 2) Asking questions, 3) Doing the questions with confidence/not relying on others, 4) Communicating and participating well in group, 5) Providing ideas/opinions, 6) Responding to the opinions of others, 7) Receiving opinion/input of others, 8) Paying attention to fellow members of other groups, and 9) Making a summary of the study. These nine aspects were assessed at the beginning of the lesson before using the WBL and were assessed after obtaining the WBL through 4 meetings. At the beginning of using the WBL, there was a low level of participant activity from various aspects measured by pre-test. Secondly, the first meeting shows an improvement in almost all aspects measured, although no significant changes have been seen yet. A sharply increased percentage can be seen in 'communicating and participating well in group'. This shows that the participants are more comfortable and happy to communicate with fellow participants because it can increase his confidence. The 'making a summary of the study' aspect is the highest aspect of the percentage. This concern is marked by group members helping each other to be able to do the finishing work. Overall, the percentage of the nine aspects measured increases after WBL

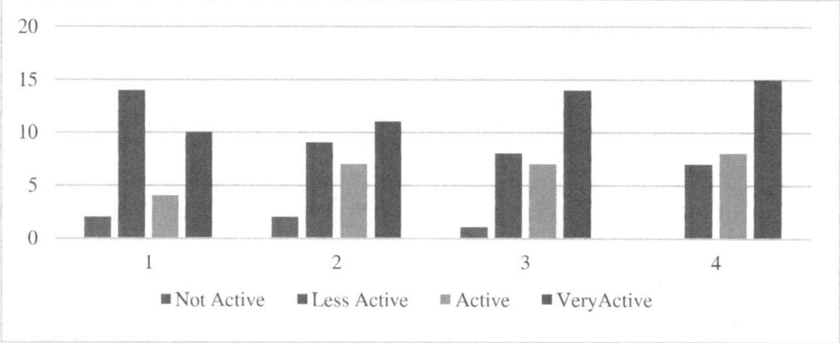

Figure 2 Total level of activity of students with WBL

Research activities with WBL conducted have shown success in the implementation. One of the advantages of WBL that experts have discovered is to increase the liveliness of participants in the learning process. Therefore, Figure 2 above shows the results of the liveliness distribution of the participants examined during the WBL execution

Building construction workers who are given work-based learning in Civil Engineering workshops are something new for the workers themselves. Generally, participants have never had a learning experience that suits the workplace. Figure 2 shows the level of activity of WBL participants observed during the learning process. The implementation of WBL is done for 4 meetings. At the beginning of the meeting, participants were still shy, afraid, with low levels of activity and participation in the learning process. Initially, the number of passive participants was 53.33% (16 persons) along with the

WBL given, the number of passive students was reduced to 3.33% (1 person). WBL participants, amounting to 30 people are a very active role in the learning process. Furthermore, with the WBL learning done there is a significant increase of each meeting. Increased participant activity is expected to have implications on the results of the final value or evaluation of the competencies of each participant.

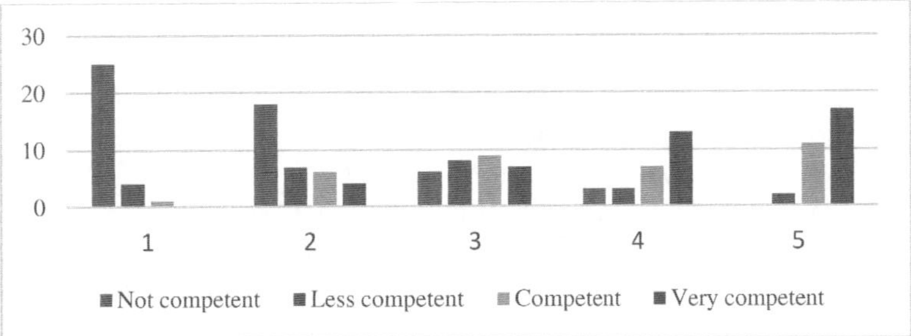

Figure 3. Recapitulation of learning outcomes participants

Figure 3 shows the evaluation of learning outcomes of the WBL process undertaken. The Y-axis is the number of WBL participants, whereas the X-axis is the learning outcome of the WBL participants. Initially, a pretest (diagram 1) was conducted to participants who produced an incompetent 96.67%. The next meetings (second, third and fourth) have been conducted by the WBL and the results can be seen in diagrams 2, 3 and 4. Along with the WBL implementation that provides increased competency of participants at the end of the meeting, there are 0% not competent. From figure 3, there is a significant increase from before doing WBL and after doing WBL

5. Discussion

The results of the study at the first meeting showed that there was a change in the participants' cognitive understanding (competence) about the plastering of the walls by applying WBL. Differences between post-test and pre-test (first meeting) were 20.59 and 25.50 (second meeting). The difference in this score does not show any major changes when it is associated with the expected competency outcomes. To improve this through the WBL, the implementation is not only done in workshops but also in the field of work in the observation or directly involved in building construction work (individuals and groups). By applying the WBL, where each student actively communicates and participates in solving a job, a cooperative understanding among the participants will be much better than individual learning. The next meeting found the difference in value between the pre-test and post-test has increased significantly, where the average final score is obtained at 84.25.

Furthermore, with the WBL in addition to obtaining the results of the participants' competency evaluation also obtained the result of evaluation activity level of the student during a learning process. Before implementing WBL, the participant's average activity was 37.72 (pre-test). While after implementing WBL the average value of activity level increased significantly from 42,22 (meeting 1), 46,67 (meeting 2), 53,05 (meeting 3) and finally increased to 67,42 (meeting 4). From the description, it can be concluded that the need for WBL to be given to participants before they enter the workforce they want.

Implementation of WBL conducted in the workshop can be continued by conducting training in the workplace. Workplace training will help construction workers about the real world of work. By implementing a guided WBL, where every participant is active in communicating and participating in finishing work, understanding will be much improved compared to without WBL. Finally, WBL can also be done not only for construction workers but for vocational education students. Providing WBL in the classroom will further enhance the competence of students who will enter the world of work.

Acknowledgments

The implementation of this research can be done with the assistance of Kemenristek Dikti RI (Ministry of Research, Technology and the Higher Education Republic of Indonesia), who has provided research grant. For that, I would like to thank for the help of material, knowledge, and guidance from related parties. I also thank the promoters and co-promoters who have provided guidance and direction so that this research can be don

References

[1] Alipour M., et al 2009, A Study of on the Job Training Effectiveness: Empirical Evidence of Iran, International Journal of Business and Management, Vol. 4, No. 11, www.ccsenet.org/journal.html download, February 27th 2017.
[2] Bailey, T.R., Hughes, K.L., & Moore, D.T., 2004, Working Knowledge Work-Based Learning and Education Reform. New York: Routledge Falmer
[3] Ball, D.L. & Forzani, F.M. 2007, What makes education research "Educational"?, Educational Researcher, 36, 9, 529-540 download November 27th 2014 dari http://er.aera.net
[4] Billet, S, Learning, 2008, Through Work: Exploring Instances Of Relational Interdependencies, International Journal Of Educational Research, Volume 47, Issue 4, 2008, pages 232-240, https://doi.org/10.1016/j.ijer.2008.07.006
[5] Bragg, D. D., Hamm, R. E., & Trinkle, K. A., 1995. Work-based learning in two-year colleges in the United States (MDS-721). Berkeley: National Center for Research in Vocational Education, University of California
[6] Djadnika, S., et al. 2005, Peningkatan Kinerja Tenaga Kerja Konstruksi Dengan Melakukan Restrukturisasi, Research Gate, DOI: 10.13140/2.1.3767.9366, https://www.researchgate.net/publication/265864692
[7] Donald P. Dingsdag, Herbert C. Biggs, Vaughn L. Sheahan, 2008, Understanding And Defining OH&S Competency For Construction Site Positions: Worker Perceptions, Safety Science, Volume 46, Issue 4, April 2008, Pages 619-633, https://doi.org/10.1016/j.ssci.2007.06.008
[8] Linehan, M. 2008, Work-Based Learning, Graduating Through The Workplace, CIT Press, Bishopstown, Cork, Ireland
[9] Little, B. et al., 2006, Employability and Work-Based Learning. London: HEA
[10] Work-based Learning Guide 2002. download February 2nd 2016, from: http://www.iowaworkforce.org/files/wlg02.pdf

Detection System of Sound Noise Level (SNL) Based on Condenser Microphone Sensor

Juniastel Rajagukguk[1], Nurdieni Eka Sari[1]

[1)]Department of Physics Universitas Negeri Medan

Corresponding author: juniastel@yahoo.com

Abstract. The research aims to know the noise level by using the Arduino Uno as data processing input from sensors and called as Sound Noise Level (SNL). The working principle of the instrument is as noise detector with the show notifications the noise level on the LCD indicator and in the audiovisual form. Noise detection using the sensor is a condenser microphone and LM 567 as IC op-amps, which are assembled so that it can detect the noise, which sounds are captured by the sensor will turn the tide of sinusoida voice became sine wave energy electricity (altering sinusoida electric current) that is able to responded to complaints by the Arduino Uno. The tool is equipped with a detector consists of a set indicator LED and sound well as the notification from the text on LCD 16*2. Work setting indicators on the condition that, if the measured noise > 75 dB then sound will beep, the red LED will light up indicating the status of the danger. If the measured value on the LCD is higher than 56 dB, sound indicator will be beep and yellow LED will be on indicating noisy. If the noise measured value <55 dB, sound indicator will be quiet indicating peaceful from noisy. From the result of the research can be explained that the SNL is capable to detecting and displaying noise level with a measuring range 50-100 dB and capable to delivering the notification noise in audiovisual.

1. Introduction

One of the undesirable condition every individual is a situation which can interfere with the comfort, e.g. hearing comfort. Condition that may interfere with the comfort of hearing is the noise. Government regulation of the Minister of the environment (1996) States, "noise is unwanted sound from the venture or activity within a certain time and levels that can cause human health problems and environmental comfort." Health problems that can be caused by noise surely will have impact, both in terms of psychological, physiological, communications and hearing. Suyatno (2010) States, "in daily life, especially the activities of work, a lot of things often spoken about comfort in the work, one of which was the disruption caused by the comfort noise that accepted. "

Humans in daily life surely hear many type of voices, like conversation, the sound of music, vehicles, and machinery, which unrealize can cause problems. Sound frequency that is too high can interfere with comport and human hearing. Many of the diseases and disorders can be caused by noise both in terms of psychological, communication and hearing. In one room, the sound is not needed will make human beings that exist within disturbed and unable to concentrate. At industry events, the sound of machine work with the frequency of the vibration exceeds the threshold of hearing will have an impact on the health of worker.

Existing noise problems of the surroundings needs to get serious attention as it can affect the balance between human and their environment. Because noise is a form of pollution generated sound. Therefore, if is not can be removed the needed a way to reduce the noise.

Noise level measurement is a test to determine how large the value of noise on an activity. Tools used to measure Sound Levels Meter (SLM) with units of Decibel (dB) in which this tool can only measure the noise level only.

The development of a more advanced technology, make human work easier by using technology. Human work easier by using technology to all forms of activities and get the result efficiently. In the case of noise in the environment, the public may know the noise levels automatically with the diverse condition of notification in accordance with the level of the frequency displayed on the output of the detector will be designed tools, so from the information the form of the anticipate the noise by using earplugs and earmuff. Based on the noise problem has been accompanied by the presence of designs a tool that relate with problem in it, which can help the field of health, industry and agencies in measuring noise levels for different purposes.

Suyatno research (2010) managed to make a noise detection instrument sound AVR microcontroller based. The Tool created has the ability measurement ± 58 dBA up to ± 95 dBA, the noise and able to provide information in three condition, that is safety, threshold danger, and very dangerousm which is indicated with a warning lamp. While research Ariyanto (2011) design tool based on noise detection library Atmega microcontroller 8535 with ouput that is already in the form of warning sounds automatically. But the result of the noise level is still on a seven segment, the sound indicated in the room was 43.4 dB. Further research of kharis (2013) makes a noise detection instrument with microcontroller Atmega 16 using a microphone as the sound catcher sensor. This tool is capable of detecting of 55 dB range value 56-75 dB with warning in the form of dot matrix LED and buzzer as a disversion. Jamaludin research (2014) makes the design and implementation of sound level meter (SLM) laboratory scale as a tool to measure the intensity of the sound-based microcontroller. This instrument is capable of measuring in relevant difference ± 5.0 % compared to SLM artificial industry. On a specific frequency value will decrease caused by external and internal conditions, the measurement should be done on a sound-proofed room.

Based on the description, we designed the noise detection instrument automatically with notification in three conditions. The condition report that displayed from instrument are safety, noisy and danger by using microphone condenser sensor with tone decoder LM 567 as amplifier circuit of sensor. Furthermore, hardware controller modules based on Arduino Uno microncontroller which contained an Atmega 328 is utilized in controlling the hardware. Data from output voltage is measured

2. Method

This work began from designing of circuits that used in SNL system. Several support devices of SNL system like circuit of power supply, noise detector, LCD display, and module voice have been developed. The construction of the system is using main components as follows microcontroller, arduino uno module, sound sensor, amplifier IC LM 567, voice module, LED and LCD.

2.1. Power Supply Circuit

The power supply circuit is used to activate the sensors on the circuit as shown at Figure 1. Transformator CT is a step down that serves to lower the voltage to 220 volt AC to 12 volt AC by using the function of the diodes on the 12 volt DC into. 12 volt DC the levelled by a capasitor so that produced 5 volt voltage regulator.

2.2. Sensor Circuit

Sensor circuit based on decoder LM 567 to detection of noise was designed and shown at Figure 2. Beside as detector, condenser microphone with the tone decoder LM 567 role as well op-amps. Operational Amplifier (op-amp) is the amplifiers has two inputs and one output voltage that is proportional to difference in voltage between the two input. The advantage of the use of operational

IOP Conf. Series: Journal of Physics: Conf. Series **970** (2018) 012025 doi:10.1088/1742-6596/970/1/012025

amplifier is ideal so that approaching characteristic in designing a circuit using this amplifier easier and be able to work on a level which is quite close to the theoretical work characteristics (Tuwaidan, 2015). In the series there are some diodes as a wave rectifier and capasitor as electric energy stronge.

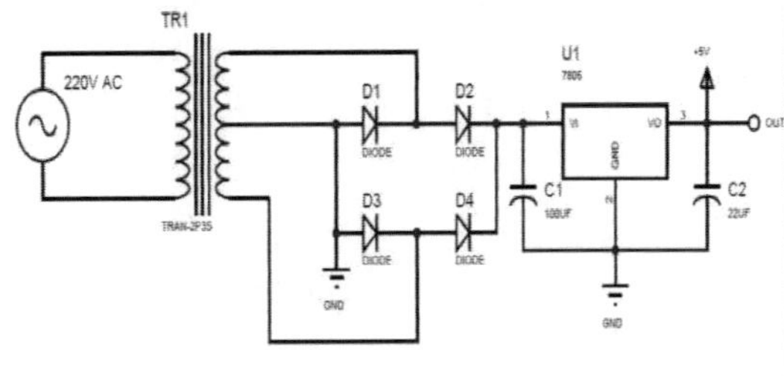

Figure 1. Power Supply Circuit

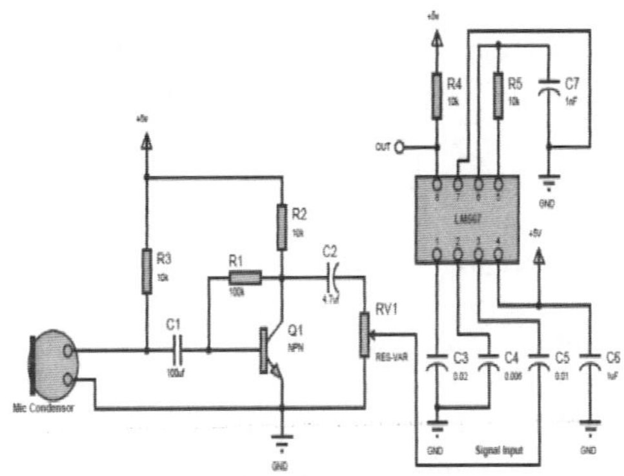

Figure 2. Noise detector sensor circuit.

2.3. LCD Circuit

LCD (Liquid crystal Display) sircuit can be seen in Figure 3. The size of LCD is 16 x 2 used as display processed data viewer microcontroller Arduino Uno. The tool is mounted outside the box series microcontroller. Configuration of 16 characters and 2 lines with each character formed by 8 pixel lines and 5 columns of pixel and 1 pixel line is krusor (Tokheim, 1994).

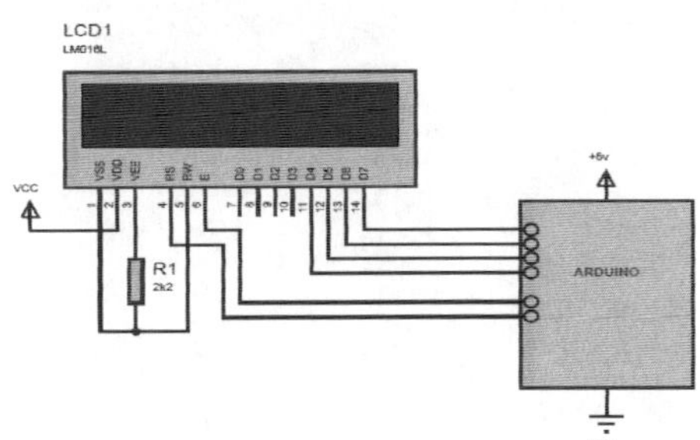

Figure 3. Series LCD which is connected with the arduino

2.4. A series of module Voice

A series of modules designed to use voice IC ISD 1508. On each type of IC ISD codes names on the last two digits indicate the maximum time duration of old that is able to be stored into the device. In this IC have include some need recording and calling data back including the microphone amplifier, sound amplifier and speaker filters to display the sound back (Tokheim,1994). In figure 4 looks to form sets of voice module.

IOP Conf. Series: Journal of Physics: Conf. Series **970** (2018) 012025 doi:10.1088/1742-6596/970/1/012025

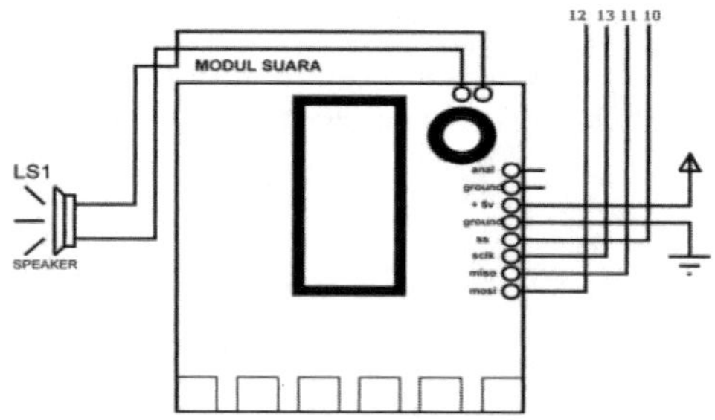

Figure 4. A series of module voice

3. Result and Discussions

3.1. Description of Noise detection Tools

Operation of the tool starting with connecting power supply to enable the detection device. The tool will display a description of the detection system noise and then measuring the noise level and displays the results of the measurements of the notification on the LCD. The design noise detector tool have some indicators include:

1. Audio Indicators:
   a. Silent is indicated safe condition
   b. Sound "don't Be Noisy" indicates the status of the noise and danger.
2. Visual indicators in the form of LED:
   a. Green LED is light indicating the safe condition.
   b. Yellow LED is light indicating the status of noise.
   c. Red LED is light indicating the status of the danger.

   These two types of indicators are working together and accompanied by a description of the notification on the LCD screen.

Figure 5. Portable object of SNL detector

3.2. Testing of Power Supply Circuits

Power supply circuit is work to serves the tension to the whole series. This series consists of two same output value amounting to 12 volts. Voltage control for micro system using voltage with a value of 5 volts. From Table 1. It can be explained that the results of testing the power supply in accordance to spesification and need 5 volt sensor supply.

Table 1. Output voltage of power supply test

No	Output Voltage (Volt)	Measured Voltage (Volt)
1	12	12
2	5	5

3.3. Testing of Noise Detection Sircuits

Calibration testing is done in a closed room, sound-proof and to avoid the existence of other sources of sound affects testing. Parameters measured from the source of the sound the buzzer this is magnitudes ADC acquired and processed as well as compared with the standard detector sound Level Meters. The results of measurements of the ADC and the comparison against the SLM standard can be seen in table 2. .

Table 2. The Value of ADC Measured By a condenser microphone and comparison with SLM standard.

No	Noise level	ADC	SLM (dB)
1	1	23	50
2	2	123	55
3	3	223	60
4	4	323	65
5	5	423	70
6	6	523	75
7	7	623	80
8	8	723	85
9	9	823	90
10	10	923	95
11	11	1023	100

Based on Table 2 it can be seen that the smallest value of the ADC sensor is 23 when the sound level is the minimum level or buzzer intensity of 50 dB. At a time when the sound level is raised in constant measurement result then the ADC also rise constantly. The big rise in the value of the ADC is of 100 for every rise in buzzer sound level. It is also directly proportional to the measuring results of the SLM that is the higher the ADC sensor value, the SLM value is also higher.

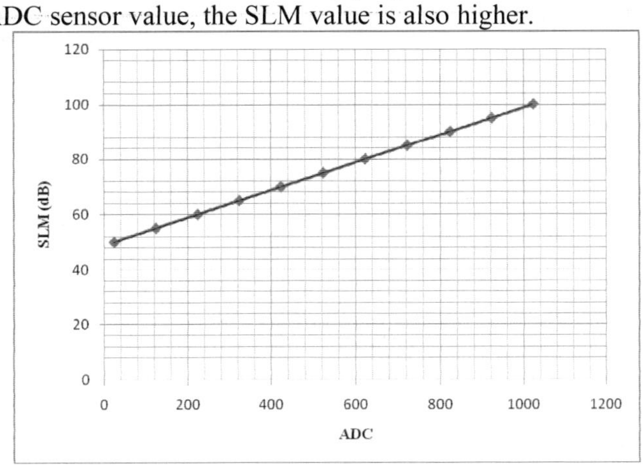

Figure 6. Noise Level Calibration Graph.

3.4. Testing of Sound Module Series

Before all of main components were combained, each

IOP Conf. Series: Journal of Physics: Conf. Series **970** (2018) 012025 doi:10.1088/1742-6596/970/1/012025

Table 3. testing voltage input sound module.

No	Condition	Voltage (Volt)
1	Beep	5
2	Mute	0

Testing voltage input modules sound on table 3 explains that when the condition voiced the voltage read on the multimeter by 5 volts while on the off voltage conditions read 0 volt. For testing the sound module series can be seen in table 4 below.

Table 4. Testing Sound Module Series

No	Sound Source (dB)	Information	Notice
1	50	peaceful	Good
2	55	noisy	Good
3	76	Danger	Good

Based on Table 4. It can be seen that for the value of the measured noise 50 dB then no sound information conveyed (silent). If the measured 55 dB noise and 76 dB sound module will pass on the information in the form of sound. From the results of the above testing can be ensured that the condition of the sound module developed in good condition and in accordance with the programming done on an arduino.

3.5. Testing of LED series

Before all of main components were combained

Table 5. the input voltage of the LED testing

No	Condition	Voltage (Volt)
1	Beep	5
2	Mute	0

Based on Table 5. The input voltage of the LED test used in this study using LED with 5 volt input voltage. After the input voltage known then testing a series of LED can be seen in table 6.

Table 6. Testing a Series LED

No	Sound Source (dB)	LED			Ket
		Red	yellow	Green	
1	50			√	Good
2	55		√		Good
3	76	√			Good

The test results Table 6 LED on the circuit when the value of the read noise 50 dB the green LED light will light up indicating the safe condition, at the time of 55 dB noise yellow LED light will turn on indicating the condition of noise and if noise had read 96 dB lights the red LED will light up indicating the condition of the dangerous. Testing the LED arrangement is said to work well and in accordance with the programming languages already designed.

3.6. Testing of SNL system

Table 7. testing tools against noise

No	Sound Level	Intensity (dB)	Status in display	Colour Indicator			Sound Indicator	
				Red	yellow	Green	Silent	sound
1	1	54	Good			√	√	
2	2	61	Noisy		√			√
3	3	66	Noisy		√			√
4	4	72	Noisy		√			√
5	5	76	Danger	√				√

6	6	81	Danger	√	√
7	7	84	Danger	√	√
8	8	91	Danger	√	√
9	9	97	Danger	√	√
10	10	100	Danger	√	√

Based on Table 7. It can be seen that the position of noise level 1 in the conditions of safe with level intensity of 54 dB. On the level of 2-5 status stated noise intensity level 61-72 dB and on the level 5-10 level stated at the intensity level danger status 76-100 dB.

CONCLUSION

Based on the research that could be taken conclusions noise detection tool-based arduino uno using sensors with tone decoder LM 567 designed worked well the percentage of errors of measurement noise of 0.7 %. With the capabilities of the tool can measure the range of the value of the 50-100 dB accompanied by notification in three conditions, that is safety, noise and danger.

References

[1] Ariyanto, Iwan, (2011), *Rancang Bangun Pendeteksi Kebisingan pada Perpustakaan UNISBANK Semarang*, Skripsi, Universitas STIKUBANK, Semarang.

[2] Jamaludin, J., Suriyanto, S., Adiansyah, D., & Sucahyo, I., (2014), Perancangan dan Implementasi Sound Level Meter (SLM) dalam Skala laboratorium sebagai Alat Ukur Intensitas Bunyi, *Jurnal Pendidikan Fisika dan Aplikasinya (JPFA)*, **Vol. 4,** No. 1 Juni 2014 42 – 46.

[3] Kharis, (2013), *Rancang Bangun Sistem Deteksi Kebisingan sebagai Media Kontrol Kenyamanan Ruangan Perpustakaan,* Skripsi, Fakultas Sains dan Teknologi, Fisika, Universitas Islam Negeri Sunan Kalijaga, Yogyakarta.

[4] Republik Indonesia, (1996), Peraturan Pemerintah Menteri Negara Lingkungan Hidup, Baku Tingkat Kebisingan, Nomor 48/MENLH/11/1996, Sekretariat Negara, Jakarta.

[5] Suyatno dan H. Ahmad., (2010), Perancangan dan Pembuatan Alat Pendeteksi Tingkat Kebisingan Bunyi Berbasis Mikrokontroler, *Jurnal Fisika dan Aplikasinya*, **Vol. 6**, No. 1 Juni 2010 : 1- 4.

[6] Tokheim, L.R., (1994), *Digital Electronic Fouth Edition*, McGRAW-HILL, Inc.

[7] Tuwaidan, Yongly. A., Eng Vecky C. Poekeol., Dringhuzen J Mamahit., (2015), Rancang Bangun Alat Ukur Desibel (dB) Meter Berbasis Mikrokontroler Arduino Uno R3, *E-journal Teknik Elektron dan Komputer*,

IOP Conf. Series: Journal of Physics: Conf. Series **970** (2018) 012026 doi:10.1088/1742-6596/970/1/012026

The Implementation of E-learning System at UIN Sumatera Utara in Response to Technology Challenge in Education

Retno Sayekti[1]

[1]Department of Library and Information Science, Universitas Islam Negeri Sumatera Utara Medan, Indonesia

*Corresponding Author: retnosayekti69@uinsu.ac.id

Abstract – The objective of this study is to find out student patterns and perceptions of using a MOODLE-based e-learning system that was first used in 2014. The methodology involved was a survey to 165 respondents comprising of several classes of various subjects. This study investigates the intensity of student's use of e-learning; time choice; justification of time choice; span of time in using e-learning; choice of access place; medium or devices used to access e-learning; and activities conducted in e-learning. In terms of students' perspectives, the author tried to find out students thought and feeling in using e-learning. Finally, the finding suggests that students tend to use various devices to access e-learning in any place that provide speed internet access. This study also revealed that most of students feel that the learning process becomes more effective and efficient by using e-learning compared to the traditional learning system in class.

1 Introduction

The development of technology has affected many aspects of human life, education is of no exception. The implementation of information and communication technology in teaching and learning has been proven to increase the effectiveness and the efficiency of the teaching and learning process. Many researches on education technology found the improvement of students' achievement in learning through the implementation of technology in teaching.

Electronic learning or e-learning is a term used to describe the action of acquiring knowledge through computer network based environments. In the last decade, the development and implementation of e-learning have become a requirement for academic institution. That is because of the advantages eLearning brings to universities.

There are many application software that have been developed to accommodate the needs of learning activities while at the same time meet teaching requirements. These applications are known as Learning Management System (LMS). One of the most popular LMS available is Moodle, an open source e-learning system which is globally used nowadays.

This study is to explore students' pattern and perception on the use of e-learning system at Universitas Islam Negeri Sumatera Utara Medan (UINSU) since its first implementation in 2014. The main purpose of the study is to improve the system that meets the needs of learning so that it can finally improve students' academic achievement.

2 Literature Review of E-Learning

Electronic learning is not a new concept in education technology. This learning design has been in existence for a while and has been used globally around the world. Many researches have been done to investigate its usefulness to improve the quality of learning process. The following description will illuminate the advantages of e-learning and software used to facilitate the learning.

IOP Conf. Series: Journal of Physics: Conf. Series **970** (2018) 012026 doi:10.1088/1742-6596/970/1/012026

2.1E-learning

E-learning has been defined as a media used for distance learning using computer technology usually Internet. [1] E-learning enables employees to learn at their work computers without travelling to the classroom. Nevertheless, the use of e-learning as media in teaching should be planned intentionally to use the network in teaching and learning in the classroom [2] E-learning also refers to anything delivered, enabled, or mediated by electronic technology for the explicit purpose of learning. The term includes: online learning, web-based learning, and computer-based training.[3]

E-learning can be used for both formal and non-formal education settings. The model can be used independently or mixed with a face to face learning activities. E-learning that used independently is merely called e-learning; while that which is mixed with a face to face learning design is always called a blended learning. The last mention concept of blended learning has been discussed in many research articles which show the effectiveness of this model. [4]–[6]

E-learning which stands for electronic learning, is a process of learning utilizing electronic devices, such as computer, laptops, notebooks, tablets, smartphones, which are connected each other through a network. The learning activities are administered by software called Learning Management System (LMS). There are many open source system for LMS which can be downloaded and develop as an e-learning facility for schools and university.

According to Bates (1995) in Zuvic-Butorac [6] e-learning has four main advantages, they are (1) enhance interactivity between students and teachers, (2) time and place flexibility, (3) potential to reach global audience, and (4) easy updating of content as well as archivable capability.

Features of e-learning includes (1) Stores and/or transmits lessons on CD-ROM, local internal or external memory, or servers on the Internet or intranet; (2) Includes content relevant to the learning objective; (3) Uses media elements such as words and pictures to deliver the content; (4) Uses instructional methods such as examples, practice, and feedback to promote learning; (5) May be instructor-led (synchronous e-learning) or designed for self- paced individual study (asynchronous e-learning); and (6) Helps learners build new knowledge and skills linked to individual learning goals or to improved organizational performance" [7]

2.2 Moodle as an E-learning Platform

In this study, Moodle is one of e-learning platform chosen to facilitate e-learning in UINSU. The consideration of choosing Moodle over other LMS software is due to its completeness in accommodating learning experience of the students and of the needs of the teachers, beside its global popularity.

Moodle has been designed to be compatible, flexible and easy to modify. [8] It was developed based on pedagogical theory of constructivism. Moodle is an Open Source Software (OSS) under GNU Public License that explicitly allow anybody to freely use, modify, redistribute and even sell the software under the condition that the open source license is maintained. This type of system has already proven very successful in developing much of the basic software that makes the Internet possible. [8] As it is an open source software, it is a challenge for teachers to design their learning activities using Moodle. Skills in technology is needed to set capability of the software to meet instructional design.

An experimental study proved that Moodle can enhance the effectiveness of learning in the following aspects: (1) better shareable materials; (2) interactive forum and chat; (3) practice and assessment quizzes; (4) gathering and reviewing assignments; (5) recording grades and report; (6) attendance; (7) glossary; (8) shout box; and, (9) face to face. [9] Another study shows that it also provides benefit to both students and teachers.[10] In the side of students, Moodle can enhance the quantity of contents, the quality of the contents, easy access, practical activities, and novelty. On the teacher side, Moodle reduces efforts and increase teaching quality.

3 Research Method

This study employs a survey method utilizing questionnaire data collection instrument. The data collection technique for this study is dependent on questionnaires administered on researcher at UINSU. Data collected are analysed and presented descriptively to elaborate students' experience in using e-learning. Respondents were instructed in the questionnaires to offer information about their experiences and impact of using E-learning. This study aims to answer the question, "What is pattern and perception of students in using e-learning of UINSU?" The questionnaire were distributed to 165 undergraduate students in the 2014/2015 academic session. The survey were to find out information about students' experiences and perception of using e-learning at UINSU.

4 The Implementation Of E-Learning At Uinsu

The implementation and development of e-learning at UINSU has begun since 2014. This transformation is supposed to produce students and human resources which are excellent in academic by providing and utilizing an electronic based learning environment. Online technologies can also be leveraged to provide a faster collaborative and interactive engagement.

The e-learning of UINSU is available and can be accessed at http://elearningiainsu.com The system was built using a Moodle version 2.6. This system enable students a lot of activities relating to their academic tasks. Below is the interface of e-learning of UIN Sumatera Utara.

Figure 1: Main Interface of E-learning UINSU

4.1 Students' Pattern of Using E-learning

This study investigates students' pattern in using e-learning from some aspects such as (1) access time; (2) place; (3) device used; (4) activities while using e-learning. The majority of students tend to access e-learning at night with the justification that it usually provides more speed internet access. They maintain that access of e-learning at campus is always slow and leads to failure of access. Nevertheless, there are still some students prefer to open e-learning at campus just before the class begin with the reason of free internet access. Students' purposeful preference of time mostly depends on the activity to be done. If they want to submit assignment or answer the online test, they would prefer to have more speed and stable internet connection. In terms of duration, most of students spend 15 – 30 minutes to do various activities in e-learning, while only few of them spend 1 – 2 hours. The duration of time used in using e-learning is determined by students' skills in using technology and tasks assigned by the teacher.

In addition to time, place is also another consideration students' take when using e-learning. They prefer to access e-learning at campus due to free internet access provided. However, there are still some students who prefer to go to Internet café (*warung Internet or Warnet*) to have more stable Internet connection. Those who consider campus as best access place maintain that in addition to free internet access, they can also discuss with their colleagues to do assignment that they need to submit into the e-learning. They access e-learning at home only if they have data package on their smartphone, or otherwise, Internet café would be their destination to go although they have to spend some rupiahs.

Most students prefer to use laptop to access e-learning (80%). This is because this device is considerably flexible in use due to its mobility capability while providing sufficient screen width to navigate in e-learning. Only is there very small number of students use their smartphone and Personal Computer (PC) to access e-learning. While smartphone becomes more affordable, those who use laptop some other time turn to their smartphone especially when they are mobile.

Students accessed e-learning for various purposes such as: (1) check syllabus or topics of the week; (2) download course resources; (3) submit assignment; (4) working on online test; (5) check grades; (6) setting their own profiles; (7) chatting with colleagues; and, (8) messaging to teachers.

4.2 Students' Perception and Expectation on Using E-learning

To find out student perception and expectation as well as their obstacles in using e-learning, open-ended questions were proposed in the questionnaire. The open-ended questions are to dig students' opinion using e-learning based on their experience. The findings are as described below.

According to students, e-learning system provides more informative media in learning compared to the conventional learning system. With e-learning system, students can access anytime to find information regarding their courses to be more precise in working on their assignment. E-learning also provides references and materials for students to refer to at their own pace of learning, as such it enables individual freedom in understanding course materials. As the objectives of the course are written in e-learning, it is clear for students the target they need to achieve at the end of the program. The online test in e-learning provides immediate feedback so that students can see their grade right after answering the questions. This, according to them, encourage transparency in grading from the teacher and encourage students' motivation in test. In short, students view e-learning as an effective media to support the process of learning. They consider a learning as a practical and flexible media of learning as well as cost effective.

Aside from its advantage, students experience some obstacles during their access to e-learning. In terms of availability of speed internet connection, it is still a big problem students face in campus. The Internet connection in campus has always been unreliable, even no connection at all in some areas. Another technical problem is database error while in use. Power off during the access, is also another big problem.

Students' expect that e-learning system implemented in UINSU should be improved its accessibility and capability. In terms of infrastructure, UINSU should provide more reliable Internet connection if the use of e-learning is to be more effective. They also suggest that all courses should be made available in e-learning.

5 Conclusion

This paper explored graduate students' pattern and perception regarding the usage of e-learning at Universitas Islam Negeri Sumatera Utara Medan, Indonesia. Literature related to the role of e-learning in improving students' learning achievements, the benefits of e-learning in teaching, as well as discussion on how the e-learning should be designed, provided the foundation for the presentation of study methods and findings. Most participants in this study have positive attitude towards the use of e-learning at UINSU and concluded that e-learning had a positive influence on their learning performance, sense of community, comfort in the online classroom and on their self-awareness. Further studies need to be done, especially on the topic of collaborative learning environment.

References

[1] A. J. Henderson, *The e-Learning Question and Answer Book: a Survival Guide for Trainers and Business Managers.* New York: American Management Association., 2003.
[2] S. Naidu, *E-Learning A Guidebook of Principles, Procedures and Practices*. 2006.
[3] D. Rossen, Elisabeth M. & Hartley, *Basics of E-learning*. .
[4] Sukarno, "Blended Learning Sebuah Alternatif Model Pembelajaran Mahasiswa Program

Sarjana (S-1) Kependidikan bagi Guru dalam Jabatan." [Online]. Available: http://download.portalgaruda.org/article.php?article=107906&val=4065. [Accessed: 14-Apr-2015].

[5] G. Sukadarmika and G. M. A. Sasmita, "Synchronization Interfaces for Improving Moodle Utilization," *Telkomnika*, vol. 10, no. 1, pp. 179–188, 2012.

[6] M. Zuvic-Butorac, N. Roncevic, D. Namcanin, and Z. Nebic, "Blended E-Learning in Higher Education: Research on Students' Perspective," *Issues Informing Sci. Inf. Technol.*, vol. 8, pp. 409–429, 2011.

[7] R. C. Clark and R. E. Mayer, *E-Learning and the Science of Instruction*. 2008.

[8] M. Dougiamas and P. Taylor, "Moodle: Using Learning Communities to Create an Open Source Course Management System," *Research.Moodle.Net*, pp. 1–16, 2003.

[9] R. Shrivastava, Y. K. Jain, and A. K. Sachan, "Designing and Developing e-Learning Solution: Study of Moodle 2.0," *Int. J. Mach. Learn. Comput.*, vol. 3, no. 3, pp. 305–308, 2013.

[10] J. Garcia *et al.*, *IT or not to be : The impact of Moodle in the education of developing countries*. 2012.

IOP Conf. Series: Journal of Physics: Conf. Series **970** (2018) 012027 doi:10.1088/1742-6596/970/1/012027

Cloud Computing Security Model with Combination of Data Encryption Standard Algorithm (DES) and Least Significant Bit (LSB)

M Basri[1*], H Mawengkang[1], E M Zamzami[1]

[1]Department of Computer Science, Faculty of Computer Science and Information Technology, University of Sumatera Utara, Jl. Universitas No. 9-A, Medan 20155, Indonesia.

[*]Email: mhdbasri@unimed.ac.id, hmawengkang@yahoo.com, elvi_zamzami@usu.ac.id

Abstract. Limitations of storage sources is one option to switch to cloud storage. Confidentiality and security of data stored on the cloud is very important. To keep up the confidentiality and security of such data can be done one of them by using cryptography techniques. Data Encryption Standard (DES) is one of the block cipher algorithms used as standard symmetric encryption algorithm. This DES will produce 8 blocks of ciphers combined into one ciphertext, but the ciphertext are weak against brute force attacks. Therefore, the last 8 block cipher will be converted into 8 random images using Least Significant Bit (LSB) algorithm which later draws the result of cipher of DES algorithm to be merged into one.

1. Introduction

Cloud computing is a metaphor of a computer / internet network that is abstracted from a concealed complex infrastructure [1] that enables one to store, change and even create applications within a virtual server owned by a cloud computing service provider. Cloud computing is basically in use for the file can be accessed anywhere and gated not in one device only. Cloud computing allows users to get access to various computing resources in a simple, periodical, efficient storage space, as well as software applications, for access to all it requires only browser applications [2]. However, some threats such as data theft known as breach / loss data [3] may occur. Therefore cloud computing as well as storing data freely but also pay attention to privacy security as the most important priority.

Attack of security loopholes using man in the middle attack (MitM) technique when sending data to cloud storage. MitM attacks are a type of attack that makes it seem that two legitimate users are communicating with each other, but there is somebody between them who sends and receives data from both users [4]. As for other attacks after obtaining user and system cloud password by using social engineering or Man-in-the-Middle Attack is by using attack on file encryption side [5]. MitM can be prevented by securing data using cryptography. However, cryptographic encryption techniques can also be attacked using cryptanalysis attacks that predict the algorithm used for message encryption. To prevent this is necessary steganography techniques.

Data Encryption Standard (DES) included into the cryptographic system symmetry and belonging to the type of block cipher. DES operates on 64-bit block size. DES encrypts 64 bits

IOP Conf. Series: Journal of Physics: Conf. Series **970** (2018) 012027 doi:10.1088/1742-6596/970/1/012027

of plaintext into 64 bits of ciphertext using 56 bits of internal key or subkey, internal keys generated from external keys 64-bits long [6]. The result of this DES algorithm will produce 8 blocks of ciphers which will then be merged into one ciphertext, but the ciphertext is weak against the brute force attack, therefore the last 8 block cipher will be converted into 8 random images using the Least Significant Bit algorithm (LSB) which later the image of DES algorithm cipher result will be merged into one.

Theoretically Least significant Bit (LSB) conceals messages secretly by replacing the data bits in the image segment with bits of secret messages. These lowest bits will then be inserted into the lowest bit or the far right bits in the pixel of the drawing composed of red, green and blue (RGB) each having an 8 bit value of 0 to 255 with binary form 00000000 to 11111111. 3 bit data can be inserted in each available image pixel. The LSB method is then applied in a library which is then named as a stagger that has a function to insert messages into images.

2. Method

Cloud security model created using PHP running localhost and online on windows operating system 10. This model will be made into the process of cloud computing where a file that will be sent to the cloud will do the encryption process so that the file will be a cipher while in the cloud storage . At the time of retrieving the file, then the file will go through the decryption process, if the key for decryption is correct then the file will be changed to plain. Illustration of the above model can be seen in the following figure 1 and 2.

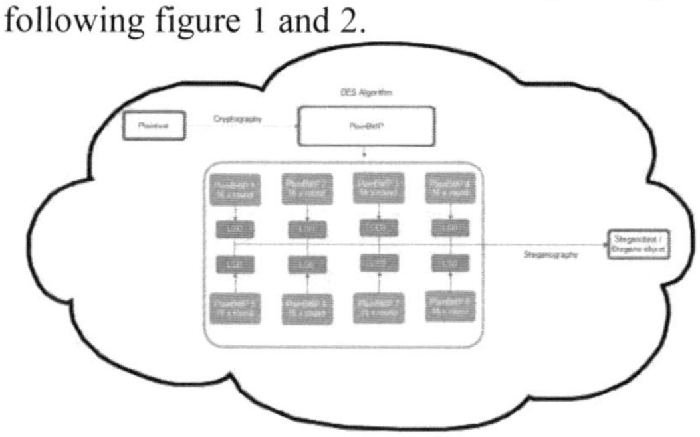

Figure 1. Design of security models.

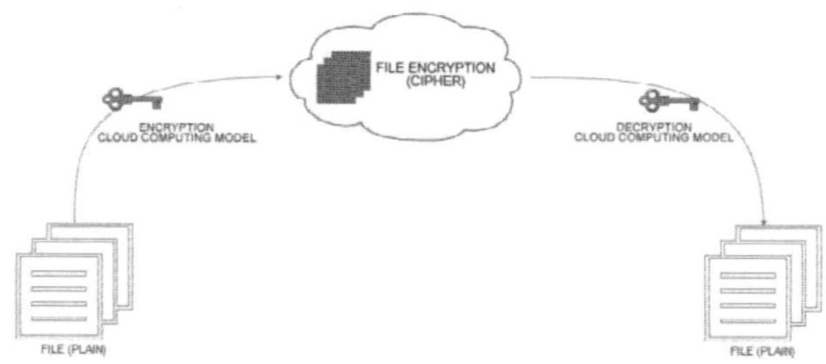

Figure 2. Design of security models in the cloud.

When the file process (figure 2) uploaded into the cloud storage file will be processed by the DES algorithm, the original file will be split into conversions into binary numbers then the result of the bits is merged into one called plaintbit and will be processed together with the bits of the key. The results of this plaintbit will then be randomized by using Initial Permutation (IP) and will go through a randomization process that will make the IP into 8 parts that each part is done 16 times the rotation process. The result of this rotation process will result in PlainBit IP.

Each PlainBit IP generated by the DES algorithm will then be processed using the LSB algorithm, then every bit processed with the LSB algorithm will be made into one in an image to be inserted, in this model the process of the DES algorithm is not completely working on the file but when the DES algorithm proceeds to the result of PlainBit IP then the result will be directed to the LSB algorithm to overcome the weakness of DES algorithm from brute force attack.

2.1. Encryption process using DES algorithm

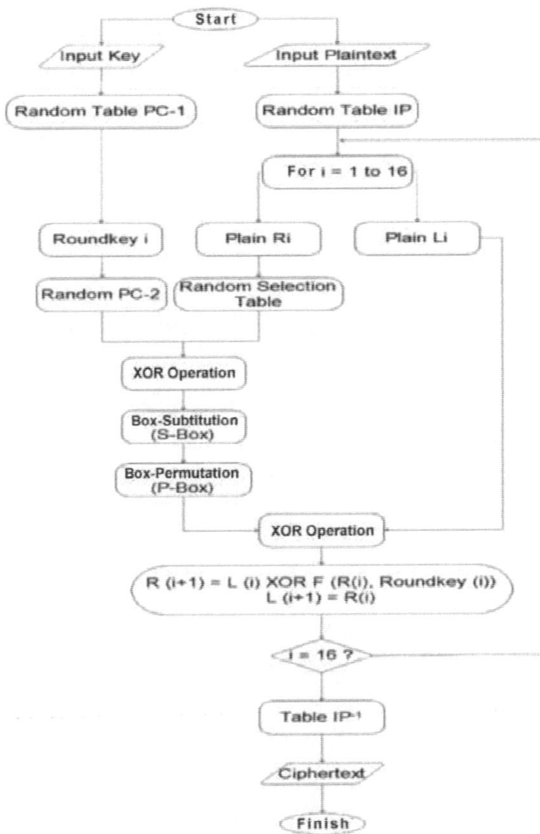

Figure 3. DES flow diagram of DES encryption process.

The process of sending text and files will be encrypted using DES 16 round algorithm to generate cipher bits from each round. DES algorithm will do the calculation of plaintext and key. The steps of the DES 16 round algorithm can be seen in figure 3.

2.2. The process of inserting bits 16 round using LSB algorithm

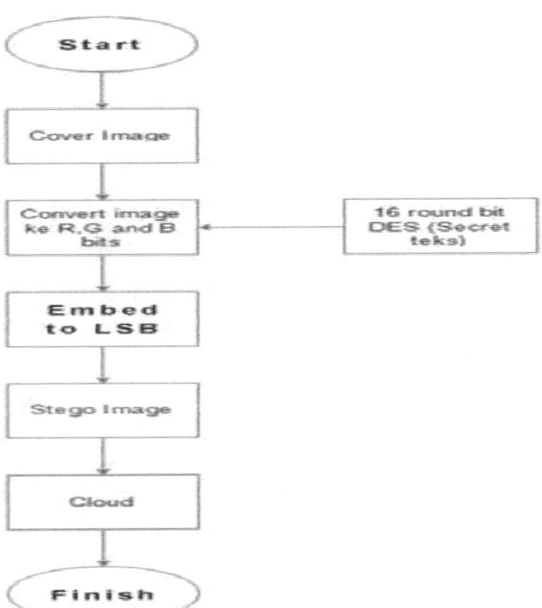

Figure 4. LSB algorithm flow diagram.

Each bit generated by 16 rounds on the DES algorithm will be inserted by using the LSB algorithm. This insertion process requires a master image / cover image. The bits will be inserted into R, G and B in the image by replacing the last bit in R, G and B so that the 16 round bits are inside the cover image. The insertion flow diagram of 16 round bits of DES can be seen in figure 4.

2.3. Decryption using LSB and DES algorithms

The decryption process is the process of returning plaintext from the cover image. Images generated by the LSB algorithm will be decrypted using keys to generate ciphertext. Ciphertext which is the output of each bit of R, G and B cover image on the LSB algorithm will generate 16-bit bits of the DES algorithm (in this process there needs to be an LSB conversion to the binner number). Chiperbit to be decrypted with bit key so it will generate cipherIPinvers, this process flow chart can be seen in figure 5 below.

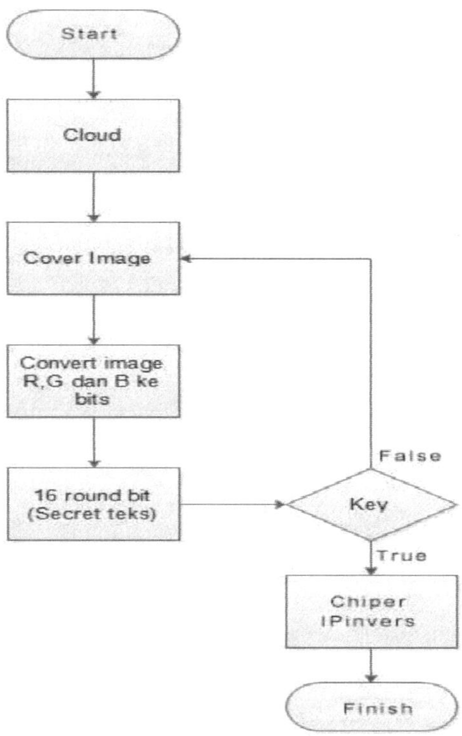

Figure 5. LSB decryption process flowchart.

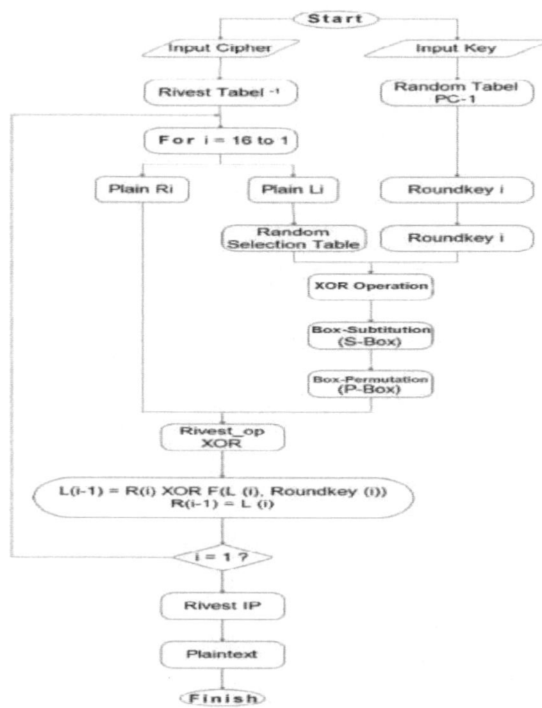

Figure 6. DES descriptor flow process diagram.

Next cipherIPinvers result from LSB decryption will be returned again into ASCII bit through 16 rounds of DES algorithm. The decryption flow diagram can be seen in Figure 6 above.

3. Result and Discussions

The results of this study is a model of the DES algorithm with a 16-round process to create a ciphertext. The 16-round process will then produce 16 parts, of which 16 parts are combined to produce a ciphertext. Ciphertext of DES algorithm has weakness in its security, as it is easy to attack cryptanalys by using brute force algorithm. Therefore in this study the DES algorithm is not processed until the end becomes ciphertext.

At the time of 16 rounds occur then each of the 16 rounds will be inserted into an image with LSB algorithm so that the cover image. The LSB algorithm allows a message to be inserted into an image by replacing the smallest bit to hide the message. To perform the implementation of this research, the authors need to conduct trials and analysis.

The display of this trial program shown in figure 7 and 8. Figure 7 shows the login form of the cloud system application and figure 8 is the main menu display after login.

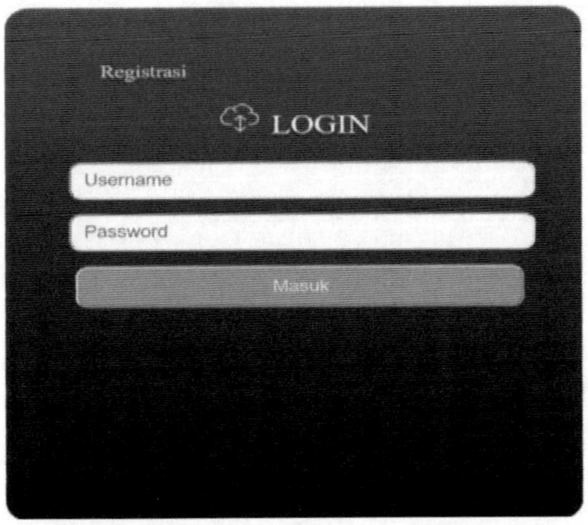

Figure 7. *Cloud* system login.

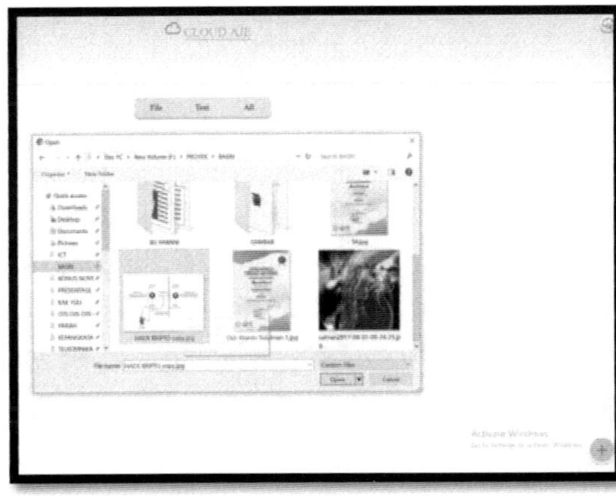

Figure 8. Form the insertion of images into the cloud.

3.1. Cloud security model of text / file encryption process

The first experiment will be the process of selecting the image of the host or the image to be inserted by the DES algorithm process using the LSB algorithm as in figure 9 above. Then select the text menu / file for the calculation process in the DES algorithm that will be inserted into the host image that has been uploaded before. Then save the encryption result in figure 10 into the host image so that this encryption result is not visible. The host image will be saved into the cloud and can not be downloaded by anyone who can not access permissions. Figure 11 is a view of the host image that has been automatically uploaded to the cloud system after passing the DES 16 round and LSB algorithm.

Figure 9. The host image to which the text will be inserted.

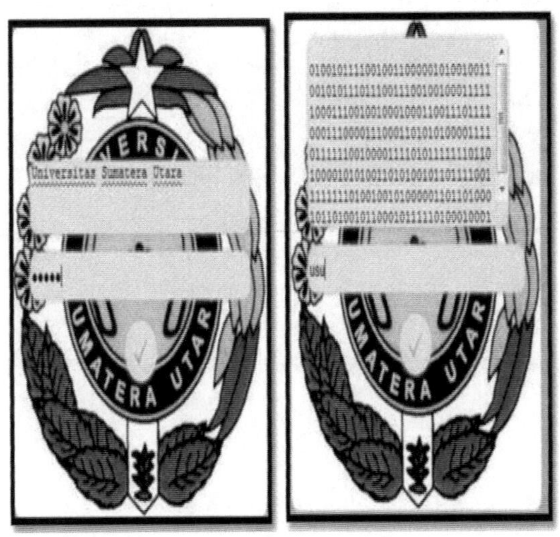

Figure 10. Text message encryption process.

IOP Conf. Series: Journal of Physics: Conf. Series **970** (2018) 012027 doi:10.1088/1742-6596/970/1/012027

Figure 11. Images that have been saved in the cloud.

3.2. Cloud security model of text / file decryption process

After going through the process of cloud computing security, the image file will be in the cloud. Users can share files that are in the cloud by sharing the link of available files, but to see the hidden messages in the image it must do the decryption first. The decryption process can be seen in figure 12. figure 13 is the process of decrypting a successful text message.

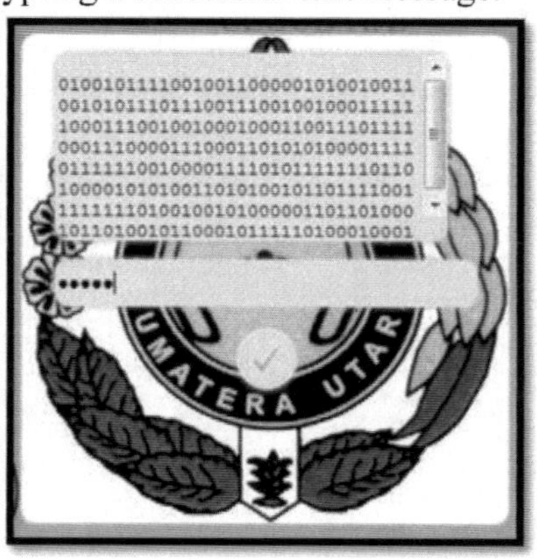

Figure 12. The image to be decrypted

Figure 13. Text message decryption process succeeded.

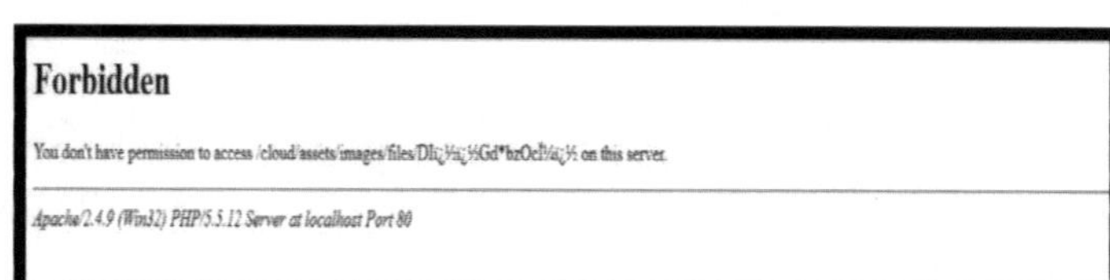

Figure 14. Text message decryption process failed

Figure 15. The file decryption process failed.

IOP Conf. Series: Journal of Physics: Conf. Series **970** (2018) 012027 doi:10.1088/1742-6596/970/1/012027

Figure 14 is a picture of a failed text message process and picture 15 is a picture on the decryption process the key entered is not the same as the key at the time of encryption then the decryption process will occur failure and the file can not be downloaded.

4. Conclusions

The conclusions of this study are as follows. First, cloud computing security model used in file and message storage processes requires the DES cryptography algorithm and LSB steganography in the 16th round of bits to keep up the security and confidentiality of files and messages. Second, process of validating files and messages that will be downloaded using the cloud computing security model if the key does not match then the file will not be downloadable.

References

[1] N A S and Hemalatha M 2012 Cloud Computing for Academic Environment *International Journal of Information and Communication Technology Research* 97-101

[2] Mathew S 2012 Implementation of Cloud Computing in Education - A Revolution *International Journal of Computer Theory and Engineering* 473-475

[3] Kaur M and Singh H 2015 A Review Of Cloud Computing Security Issues *International Journal of Advances in Engineering & Technology (IJAET)* 397-403

[4] Gaikwad B P 2014 A Critical Review on Risk of Cloud Computing in Commercial *International Journal of Innovative Research in Computer and Communication Engineering* 1-8

[5] Ahmed E S and A Saeed R 2014 A Survey of Big Data Cloud Computing Security *International Journal of Computer Science and Software Engineering (IJCSSE)* 78-85

[6] Kaur N and Sodhi S 2016 Data Encryption Standard Algorithm (DES) for Secure Data Transmission *International Conference on Advances in Emerging Technology (ICAET)* 31-37

Development of Mathematics Learning Strategy Module, Based on Higher Order Thinking Skill (Hots) To Improve Mathematic Communication And Self Efficacy On Students Mathematics Department

Ade Andriani[1], Izwita Dewi, Budi Halomoan

Universitas Negeri Medan

Corresponding author: andrianiade@ymail.com

Abstract. In general, this research is conducted to improve the quality of lectures on mathematics learning strategy in Mathematics Department. The specific objective of this research is to develop learning instrument of mathematics learning strategy based on Higher Order Thinking Skill (HOTS) that can be used to improve mathematical communication and self efficacy of mathematics education students. The type of research is development research (Research & Development), where this research aims to develop a new product or improve the product that has been made. This development research refers to the four-D Model, which consists of four stages: defining, designing, developing, and disseminating. The instrument of this research is the validation sheet and the student response sheet of the instrument.

1. Introduction

In an effort to qualify for university graduates in Indonesia, the government has issued Presidential Regulation No.08 of 2012 on the Indonesian National Qualification Framework (KKNI). Given this KKNI is expected to change the way one sees one's competence, no longer merely a diploma but by looking at a nationally agreed upon qualification framework as a basis for the recognition of a broad and accountable (formal, informal, or formal) outcome of a person's education. Along with the pattern of changes that occur due to changing patterns of graduate requirements and curriculum changes based on KKNI and Implementation of the 2013 curriculum,

According to Saragih (2015) most teachers teach mathematics by explaining the concepts and operations of mathematics, giving examples of doing things, a little questioning (if any), followed by asking students to work on similar problems with the problems already explained by teachers. The researcher's experience when becoming a PLPG instructor saw that many teachers who teach with lectures or learning activities are still teacher-centered. It certainly boils down to the teacher's "kitchen", perhaps when they are still studying with learning resources that do not support the learning process that is centered on the students. When students study with learning tools that allow them to develop higher order thinking, they are likely to become qualified teachers.

During this time the lecturers of mathematics teaching strategies in mathematics majors using different learning sources or modules. For that reason researchers want to develop a module of mathematics learning strategy based on higher order thinking skills to improve mathematical communication and self efficacy students in math majors.

2. Review of literature and definition of terms

2.1 Higher Order Thinking Skills

High-order thinking or commonly known as higher order thingking (HOT) is an important factor in mathematics learning. This is evident from what is well-documented in the Education Unit Level Curriculum (KTSP) as well as the National Council of Teachers of Mathematics (2000), mathematics education is aimed at the development of skills aspects (Puskur, 2005):.

1. The ability to mathematical problem solving, other lessons, and real-life problems
2. The ability to use mathematics as a communication tool (*mathematical communication*)
3. The ability to connect mathematical ideas (*mathematical connections*);
4. The ability to reasoning (mathematical reasoning) that can be used in any situation, such as critical thinking, logical, and systematic; be objective, honest, discipline and solve problems.
5. Positive attitudes toward mathematics.

Thomas dan Thorn (2009) say *HOT takes thinking to higher levels than restating the facts and requires students to do something with the facts — understand them, infer from them, connect them to other facts and concepts, categorize them, manipulate them, put them together in new or novel ways, and apply them as we seek new solutions to new problems.*

2.2 Mathematical Communication Skills

In addition to thinking tools, mathematics as well as language. Mathematical language also has the same function as common language, that is (1) naming or labeling, (2) interaction and (3) information transmission. In expressing mathematics, symbols are used to represent objects, share ideas, and share information. But the language of mathematics must also have the character of mathematics, using logic and words used as agreed or clearly defined.

Dewi (2009) says mathematical understanding can be related to mathematics as a language, symbol language, which is seen in symbolization and formulation that is transforming a statement into formulas, symbols or drawings. The relationship between language and mathematics as a symbol language is very close. Symbol language is a tool to explain a mathematical concept to be simple. In contrast, mathematics is a tool to simplify one's math understanding into symbol language.

Mathematics as a language called "the language of science" as well as the language of engineering and language economy (commerce). Even mathematics is referred to as the "universal language," therefore humans all over the world can use mathematics to communicate even though they have different languages. Then Baroody (1993) also says there are two reasons why communication is needed in mathematics learning. (1) Mathematics as a language means mathematics not only as a tool of thought, finding patterns, or solving problems but also as a means of communicating ideas clearly, precisely and concisely; and (2) mathematics as social activity, meaning as language, mathematics is a social activity in learning, such as the interaction between students with teachers, and student interaction with students

A person's mathematical understanding can not be seen and heard, but mathematical communication facilitates students "voicing" what they think about their mathematical knowledge. The explanation implies that mathematical learning that gives students the opportunity to communicate mathematically can make students feel confident and have a positive attitude

2.3 Defenition of Self Efficacy

Walgito (2011) defines self efficacy as one of the most influential aspects of self-knowledge in everyday human life. This is due to the efficacy of self-owned influence the individual in determining the actions to be taken to achieve the objectives including the estimated incidents to be faced.

The basic concept of self efficacy theory is the problem of the belief that every individual has the ability to control his thoughts, feelings and behavior. Thus self efficacy is a matter of subjective

IOP Conf. Series: Journal of Physics: Conf. Series **970** (2018) 012028 doi:10.1088/1742-6596/970/1/012028

perception. Self efficacy is defined as one's consideration of his or her ability to achieve desired or determined levels of performance, which will affect the next action. According to Zulkosky (2012) says that self-efficacy also affects people's choices in behavioral settings, the number of their efforts to complete the task, and the length of time they stay in the face of obstacles. .

3. Methodology

This research is a research development (Research and Development) is a type of research that develops a new product or refine a product that has been there before. This development research refers to the development of 4-D (four-D Model) proposed by Thiagarajan, which consists of four stages: define, design, development, disseminate. This development model has been focused on the development of module. Module that are developed is a book of mathematics learning strategy..

Analysis of validators in the form of values that have a range and descriptive value of suggestions and comments. Determining the average value of the expert for each indicator by the formula:

$$Average(Va) = \frac{\text{Total Score}}{\text{number of aspects}}$$

Va : validation value

For student response data is analyzed by counting the percentage of many students giving positive response and on each category stated in the questionnaire by using the following formula:

$$\text{Percentage} = \frac{\text{A}}{\text{B}} x100\%$$

A : The proportion of students who choose
B : Number of students (respondents)

The student's response is said to be positive if 80% or more of the students respond in a positive category (happy, new, interested, clear, and interested) to every aspect that is responded.

The effectiveness of module is seen from the data of student activity improvement with the formula:

$$\text{Percentage} = \frac{\text{the frequency that appears}}{\text{Total Frequency}} x100\%$$

The increased value of mathematical communication is analyzed by the formula

$$Indeks\ Gain = \frac{\text{Postes} - \text{Pretes}}{\text{Ideal Score - Pretes}}$$

While the improvement of self efficacy is seen in each trial and between trials. improvement seen from the change of self efficacy indicator in a better direction.

4.Results and Discussion

Having analyzed the validation sheet of experts, then obtained a count of validity and that is

Table 1. Validation Results

Validator	Value	criteria
Validator I	4,9	Valid
Validator II	4,76	Valid
Validator III	4,85	Valid
Average	4,83	Valid

Based on table 1 it can be concluded that the module developed is valid and can be continued testing process in small group, while comments in the form of suggestions and criticism used to revise module

4.1 Small Group Anlisis

Small group trial was conducted to 10 students who have taken courses in mathematics learning strategies, as a test of the legibility of learning materials that are developed that is a mathematics learning strategy book.Here is the table of test results of the legibility of module.

Table 2. Module readability

	Score	Pecentage	criteria
Average	41.36	83.2	Baik

From the results seen in Table 2, it is illustrative that the module of mathematics learning strategy has a good readability value or in other words it can be concluded that the book of mathematics learning strategy is easy to understand.

4.2 Large Group Trial Analysis

The big group is here is a semester V students majoring in mathematics education FMIPA UNIMED who took the mathematics learning strategy, which amounted to 31 students, this analysis to see the practicality and effectiveness of the module

Table 3. Practicality of Module

	Score	Percentage	criteria
Average	21,12	84.5	Baik

From Table 3, the test value of practicability is at 84.5 with good criteria, in other words a practical mathematics learning strategy book to use.

Table 4. Improvement Of Math Communication Skills

	Pre-test	Post-test	Gain
Average	50.8	81.2	0.61

Based on Table 4 above shows an increase of the average score of students' mathematical communication ability from pre test to post test. Based on the increase index is 0.61 which is categorized as high or significant increase.

Table 5. Student Response

	Positive	Percentage	Negative	Percentage
Average	26.8	86.45	4.2	13.5

Based on Table 5 above, the average that responded positively was 86.45% while the negative responded 13.5%

Table 6. Student Activity

	Pertemuan						Average
	I	II	III	IV	V	VI	
Student Activity	83.3 %	82.6%	84.6%	82.6%	81.3%	92%	84.4%

Based on Table 6 above, the average student activity for 6 meetings is 84.4% with good category.

4.3 Discussion

4.3.1.Define The stage to determine the conditions of learning that begins with curriculum nanalisis, curriculum used by this students in the Faculty of Mathematics and Natural Sciences of Medan State University is a block-based curriculum. Then analyzing the character of the students, semester students of mathematics education is not familiar with the subjects of mathematics learning strategy, because each course in the previous semester is integrated with some mathematics learning strategies, such as approach and learning model. Cognitive students are already at the operational level of formal meaning already able to think abstract. In the next stage the researcher identifies the material on mathematics learning strategies, collects and selects relevant material and organizes systematically. The result of syllabus and material analysis then formulated the purpose of the lecture and its indicator in accordance with the competency standard of the mathematics learning strategy.

4.3.2. Design This stage the researchers designed the prototype module begins with the preparation of the test. The test is used to measure students' mathematical communication skills, the test is a matter consisting of 4 items that will be tested before and after the learning. In addition to the test also prepared a questionnaire that is used to measure the criteria of student self efficacy. Next select the media, media used in mathematics learning strategy in the form of book media. The reason for choosing a book media is that students easily analyze, search for information in books, solve problems, make conclusions about mathematics learning strategies and applications in education such as schools. Then make the first design in the form of prototype which will be validated to the expert validator.

4.3.3. Development This stage begins with the testing of draft 1 to three lecturers from state university of medan. A good criterion of a book if the validity score is in the range of 4-5. From table 1 the average mean value given by the expert is 4.83 which gives the illustration that the module is valid and can be continued to the small group test, this test is done after revising the book according to experts' comments

After being reviewed by an expert and revising the book produced draft 2 which will be test of readability.This tested to 10 mathematics students who have taken the course of mathematics learning strategy, this test was conducted on August 15, 2017. Based on table 2, the overall average is 83.2% with good category or this module easy to understand. Next is a large group test, which is to test the effectiveness and practicality of teaching materials by testing the module's implementation, student activities, student responses and improvement of communication skills and self-efficacy enhancement. This test is done to 31 students who are taking courses in mathematics learning strategy. The test results are used to revise the draft 2 so as to produce draft 3. From table 3 can be seen the overall average of 84.5% with good practicality category. Textbooks developed can be used practically in the learning process of mathematics learning strategy. Table 4 shows the improvement of students' mathematical communication skills. In the first meeting of the lecture on August 24, 2017, the first test is pre test (pre test), the average pre test of the students is 50.8 is still in the category of E score or not pass, then at the 7th meeting of the lecture done the final test (post-test) on October 10, 2017, the average post test score is 81.2 with category B value (pass). There was a significant increase by seeing the average value of gain is 0.61 with the category of high increase.

Similarly, the value of student self efficacy, self efficacy questionnaire distributed at the beginning and the meeting to 7 lectures. At the initial meeting the results of self efficacy test already has a good category, after 6 lecture meetings self efficacy test results increased towards the positive.
The use of mathematics learning strategy book gives good result on student activity in learning. Seen in table 7, the average student activity for 6 meetings was 84.4% with good activity category. Similarly, the response given by students to the book is equal to 86.4% which gives a positive response.

IOP Conf. Series: Journal of Physics: Conf. Series **970** (2018) 012028 doi:10.1088/1742-6596/970/1/012028

5.Conclusion

The process of developing the module in the form of a book of mathematics learning strategy is done up to stage 3 that is define, design and development. The design phase produces draft 1, then in the development phase, draft 1 which is validated by 3 validators and revised according to expert comments then produce draft 2, then draft 2 tested small group that yields good legibility value. Furthermore the book tested large groups produce value of practicality and good effectiveness.

6. Refferences

[1] Baroody,A.j.(1993). *Problem Solving, Reasoning and Communicating, K-8(helping children think mathematically.*Newyork: Macmillan Publishing Company

[2] Dewi, Izwita. (2009). *Profil Komunikasi Matematika Mahasiswa Calon Guru ditinjau dari Perbedaan Jenis Kelamin.* Disertasi Tesis tidak dipublikasikan. Program Pascasarjana Surabaya : Surabaya.

[3] National Council of Teacher of Mathematics. (2000). *Principles and Standards for School Mathematics*. Reston, VA: NCTM.

[4] Puskur (2005). *Kurikulum dan Hasil Belajar. Kompetensi Dasar Mata Pelajaran Matematika Sekolah Menengah Pertama dan Madrasah Tsanawiyah.*Jakarta: Balitbang Depdiknas.

[5] Saragih., S. (2015) *Pengembangan Model Pembelajaran Berpusat pada Siswa untuk Meningkatkan Kemampuan Berpikir Matematika Tingkat Tinggi*. Laporan Penelitian Hibah Pasca Tahun I, Unimed. Medan.

[6] Thomas, A., and Thorn, G. (2009) How to Increase Hihger Order Thingking.http://www.readingrockets.org/articles/by-author/59994

[7] Walgito, Bimo. 2011. *Teori-teori Psikologi Sosial.*Yogyakarta : C.V Andi Offset.

Biology Factual Knowledge at Eleventh Grade of Senior High School Students in Pacitan based on Favorite Schools

I A Yustiana[1], Paidi[2], I S Mercuriani[3]

[1]Post-graduate Students of Biology Education, Yogyakarta State University, Yogyakarta, Indonesia
[2,3]Biology Education Department, Yogyakarta State University, Yogyakarta, Indonesia

[1]E-mail: aidayustiana@gmail.com

This study aimed to determine the Biology factual knowledge at eleventh grade of senior high school students in Pacitan based on favorite schools. This research was a descriptive research by using survey method. The population in this study was all of senior high school students in Pacitan. The sampling technique used purposive sampling technique and obtained 3 favorite schools and 3 non-favorite schools. The technique of collecting data used test form which was as the instrument of the research. Data analysis technique used Mann-Whitney U test. Based on the test, it was obtained p = 0,000 (p <0,05) so there was a significant difference between the factual knowledge of the students in the favorite schools and non-favorite schools in Pacitan. The factual knowledge of students in favorite schools was higher with an average of 5.32 while non-favorite schools were obtained an average of 4.36.

Keywords: Factual Knowledge, Biology, Senior High School Students

1. Introduction

One aspect of competence that has been emphasized in the curriculum compilation in Indonesia both *Kurikulum Tingkat Satuan Pendidikan (KTSP)* and Curriculum 2013 is the development of students' thinking ability. Permendikbud No.23 of 2006 on Graduate Competency Standards at SMA / MA / SMALB education level states that through the learning process students are expected to build and apply information and knowledge in a logical, critical, creative, and innovative and demonstrate the ability to think logical, critical, creative, and innovative in making the decision.

Graduate Competency Standards Curriculum 2013 or *Standar Lulusan Kurikulum* 2013 Permendikbud Number 54 of 2013 states every graduate elementary and secondary education units have competencies in three dimensions of attitude, knowledge, and skills. In the aspect of knowledge or cognitive domain students are expected to have factual, conceptual, procedural, and metacognitive knowledge in science, technology, art, and culture with the insights of humanity, nationality, state and civilization on the causes and effects of phenomena and events.

In line with the regulation in 2016 through Permendikud No.20 Year 2016 about Competency Standards Graduates of Primary and Secondary Education firmly state that the need for high school students to be trained in cognitive processes that cover the six process categories (C1-C6) from recalling, understanding, applying, analyzing, and evaluating to achieve the mastery of factual, conceptual, procedural, and metacognitive knowledge.

This is in accordance with the cognitive theory of Benjamine Samuel Bloom which in its development, Bloom's taxonomy was revised by Anderson and Krathwohl. Bloom's Taxonomy has classified learning outcomes in three domains: cognitive, affective, and psychomotor. In the cognitive domain includes six categories C1-C6 that is remembering, understanding, applying, analyzing, evaluating, and creating. While the dimensions of cognitive products consist of four categorical such as factual knowledge, conceptual knowledge, procedural knowledge and knowledge of metacognition.

The level of favorite school is likely to affect students' achievement in which the school favorability level is one of the indicators related to students' achievement in the school. This is in accordance with the statement of Mutiara (2005) that indicators of school favorites from the community, among others (1) public interest to enter the school is high (2) achievements obtained by many schools (3) have adequate infrastructure (4) good and high graduate scores (5) the number of learners received at favorite schools or colleges. Indicators from *Dinas Pendidikan* are more or less the same as the community's assumptions, among others (1) the school has the infrastructure that has fulfilled *Badan Standar Nasional Pendidikan* (BSNP) (2) have many achievements both academic and non-academic (3) produce good and excellent graduates 4) the number of students received at a favorite college.

Until now there is no comprehensive information on the implementation of learning that accommodates these cognitive processes and products in the schools in Pacitan regency. One of them is how factual knowledge of students is viewed from the favorite schools, especially through Biology subjects. This factual knowledge is a fundamental knowledge that students must have when it comes to learn or solve the problems within a discipline. Given this knowledge is the basic capital of learners, so it is important to know the level of its achievement.

Therefore, it is needed the research under the title "Biology Factual Knowledge at Eleventh Grade of Senior High School Students in Pacitan based on Favoritism School". This study is expected to provide information about the achievement of factual knowledge of learners in Pacitan regency so that the teachers can design a more effective and efficient learning based on the results of the analysis that has been done.

2. Materials and Methods

2.1 Materials

2.1.1 Factual Knowledge

Factual knowledge is the knowledge in the form of pieces of information or basic elements that exist in a particular discipline. Factual knowledge is generally a low-level abstraction. There are two kinds of factual knowledge, knowledge of terminology and knowledge of specific details and elements (Widodo, 2006).

Knowledge of terminology: includes knowledge about certain labels or symbols both verbal and non-verbal. Each discipline usually has specific terminology for the discipline. Some examples of knowledge about terminology are knowledge of the alphabet, knowledge of scientific terms, and knowledge of symbols in maps.

Knowledge of specific details and elements: includes knowledge of events, persons, times and other information of a very specific nature. Some examples of knowledge about details and elements, such as knowledge of names, place and times of events, knowledge of a country's products, and knowledge of information resources. Due to the immense amount of facts, educators need to select and sort out which facts are very important and which facts are less important (Anderson, et.al, 2001).

2.1.2 Favoritism School

School favorability in this case refers to superior and effective schools. Indicators of favorite schools from the community, among others, the public interest to enter the school is high, the achievements obtained by many schools, have adequate infrastructure, good graduates and high graduate scores, and the number of students received at favorite schools or colleges (Pearl, 2005).

In line with these indicators, Alzahrani (2016) stated that good schools can be identified based on alignment of organizational structure to school mission and goals, integration and effectiveness of management processes (resource allocation, budgeting, planning, performance evaluation, etc.), effective and valuable school facilities and utilities, effective and valuable human resource processes, knowledge sharing, inclusive working and learning environment, employee development and empowerment, and performance review and evaluation.

2.2 Methods

2.2.1 Research Design

This research was descriptive research using survey method. This research was conducted in Pacitan regency of East Java province. The data collection was conducted on May 4th to May 13th, 2017, the even semester of the academic year 2016/2017, in accordance with the academic calendar of the school, related to the subject matter used in the study. The sample in this research is the eleventh grade students in Pacitan regency taken from 3 favorite schools and 3 non-favorite schools with the total students are 309 students.

2.2.2 Research Instrument

The data were collected through the test technique. The test technique is performed to measure the achievement of factual knowledge. The test instrument used was a question instrument developed by Paidi, et.al (2016) which consists of immune system material, reproductive system, coordination system, and excretion system.

2.2.3 Data Analysis

The collected and validated data are analyzed descriptively and inferentially. Descriptive analysis to find the maximum score, minimum score, average student score, standard deviation, range and variant. Inferential analysis was conducted to know the difference of achievement of factual knowledge of students in school with variable that accommodated, that is the favorability school. The significance of factual achievement figures is associated with the criteria presented in table 1 below.

Table 1. Range Achievement Factual Knowledge in Biology Subject

Range of Mean	Achievement Category of Factual Knowledge
$(7.5) \leq M \leq (10)$	Very High
$(5) \leq M \leq (7.5)$	High
$(2.5) \leq M \leq (5)$	Low
$(0) \leq M \leq (2.5)$	Very Low

3. Result and Discussion

3.1 Result

The achievement of students' factual abilities, shown from the average, standard deviation, minimum value, and maximum calculated from all sample members, are presented in table 2 below.

Table 2. Mean and Standard Deviation Factual Knowledge

Statistics	Score	Category
Mean	5.07	High
STDEV	0,05	-
Minimum Score	1.06	-
Maximum Score	7.46	-

The table shows that the factual knowledge of the students at the eleventh grade of IPA in Pacitan regency is 5.07 which belong to the high category. From a very wide range of minimum and maximum values, it is equal to 6.4 (7.46-1.06) with a standard deviation of 0.05.

Table 3. Factual Knowledge at the Levels of Favorability of Different School

Variable	Statistics (Mean)	Criteria	Significance (Mann-Whitney Test)
Favorite	5.32	High	Significance Different (p=0,000)
Non-Favorite	4.36	Low	

Associated with the schools' favorites level in table 3, it is seen that there is a significant difference in the factual knowledge of students between favorite and non-favorite schools with p = 0,000 (p <0.05). Schools that include favorites get an average of 5.32 while non-favorite schools are 4.36.

3.2 Discussion

The factual knowledge of students in the Pacitan regency included in the high category so that students were expected to build higher knowledge because they already have a good foundation or capital. However, in relation to the accommodated variable of favorite school, it was found that there was a significant difference between schools that included favorites and non-favorites. Favorite schools tend to have higher factual knowledge while non-favorite schools are still low. According to Aischa (2010), favorite schools graduates are better than non-favorite schools.

This is reinforced by Dionisia (2016), that the thinking ability of divergent students in favorite schools is higher than non-favorite schools. In response, teachers should pay more attention for developing students' factual abilities if they want to achieve more complex abilities. One way is by choosing methods and learning models that emphasize the facts. This factual knowledge is a fundamental knowledge that students must have when it comes to learn or solve the problems within a discipline.

4. Conclusion

From the results of this study, it can be concluded that the factual ability of students in Pacitan regency including high criteria and students' factual ability is significantly influenced by the level of school favorability.

References

[1] Aischa. 2010. *Memilih Sekolah untuk Anak*. Bandung: Inti Medina

[2] Alzahrani, Saeed M.H. 2016. *Identifying Characteristics of a "Good School" in the British and Saudi Arabian Education Systems*. Journal of Education and Practice **Vol.7**, No.27, (Manchester Metropolitan University, United Kingdom) p 146

[3] Anderson L.R, Krathwohl D.R, et al. 2001. *A Taxonomy for Learning, Teaching, and Assessing: A Revision of Bloom's Taxonomy of Educational Objectives*. A Bridged Edition. New York: Longman.

[4] Dionisia. 2016. *Kemampuan Berpikir Divergen Keterampilan Proses Sains Aspek Biologi Siswa Kelas V Sekolah Dasar di Kota Yogyakarta Ditinjau berdasarkan Kefavoritan Sekolah.* Yogyakarta: FMIPA UNY

[5] Mutiara, Ipa Ratna. 2005. *Perbedaan Sekolah Favorit dan Sekolah Non Favorit di Medan*. Tesis. Medan: Universitas Negeri Medan.

[6] Permendikbud. 2006. *Peraturan Menteri Pendidikan dan Kebudayaan Nomor 23 Tahun 2006, tentang Standar Kompetensi Lulusan (SKL) SMA*.

[7] Permendikbud. 2013. *Peraturan Menteri Pendidikan dan Kebudayaan Nomor 54 Tahun 2013, tentang Standar Kompetensi Lulusan (SKL) SMA*.

IOP Conf. Series: Journal of Physics: Conf. Series **970** (2018) 012029 doi:10.1088/1742-6596/970/1/012029

[8] Permendikbud. 2016. *Peraturan Menteri Pendidikan dan Kebudayaan Nomor 20 Tahun 2016, tentang Standar Kompetensi Lulusan (SKL) SMA*.
[9] Widodo, A. 2006. *Revisi Taksonomi Bloom dan Pengembangan Butir Soal*. Buletin Puspendik. 3(2), p 18-29.

Implementation of Data Mining to Analyze Drug Cases Using C4.5 Decision Tree

Sri Wahyuni

Lecturer of Computer Engineering in University of Pembangunan Panca Budi

Corresponding author: sriwahyuni@dosen.pancabudi.ac.id

Abstract. Data mining was the process of finding useful information from a large set of databases. One of the existing techniques in data mining was classification. The method used was decision tree method and algorithm used was C4.5 algorithm. The decision tree method was a method that transformed a very large fact into a decision tree which was presenting the rules. Decision tree method was useful for exploring data, as well as finding a hidden relationship between a number of potential input variables with a target variable. The decision tree of the C4.5 algorithm was constructed with several stages including the selection of attributes as roots, created a branch for each value and divided the case into the branch. These stages would be repeated for each branch until all the cases on the branch had the same class. From the solution of the decision tree there would be some rules of a case. In this case the researcher classified the data of prisoners at Labuhan Deli prison to know the factors of detainees committing criminal acts of drugs. By applying this C4.5 algorithm, then the knowledge was obtained as information to minimize the criminal acts of drugs. From the findings of the research, it was found that the most influential factor of the detainee committed the criminal act of drugs was from the address variable.

1. Introduction

Data Mining is one of the fastest growing fields due to the huge demands for added value of large-scale databases that are in line with the growth of information technology. Data mining can extract large data sets into new knowledge.

Classification is a process of finding models that explain or distinguish concepts or classes of data, in order to be able to predict the class of an object whose class is unknown [1]. In the Classification there are several methods namely decision tree, neural network, rought set theory, Bayesian theory and fuzzy logic [2]. And each function has its own algorithm. C4.5 is a well-known and used for data classification that has numerical and categorical attributes. In previous research, Angga and Riani (2012) used data mining to predict the criteria of customers credit using the classification method used on a decision tree which later result in an application that allows the bank's funds to obtain credit marketing targets [3]. Muhammad Syahril (2011) used data mining to convert hypertension disease training data [4]. Sri wahyuni (2017) also uses data mining to analyze dropped- out students data. This research uses Decision Tree C4.5 algorithm that belong in Classification function. The decision tree method turns a very large fact into a decision tree that represents rules. The purpose of this study is to analyze the data of prisoners in Labuhan Deli prison extracted with data mining decision tree algorithm C4.5, then it will generate new knowledge to learn about the factors of detainees doing drug crimes.

2. Literature Review

1. Understanding Data Mining

In a book titled "Data Mining", data mining contains a searching process of trend or pattern in a large database for future decision making. It is expected that data mining devices are able to recognize these patterns of data with minimum input. These patterns are recognized by certain devices that can provide a useful and insightful data analysis that can later be studied more closely, which may use other decision-taking support tools [5]. Data mining deals with a sub-area of statistics called exploratory data analysis, which has the same objective and relies on statistical measures. Data mining is also closely related to the artificial intelligent sub-area called knowledge discovery and machine learning. An important characteristic of data mining is that the data volume is huge even though the ideas from the related study area can be applied to data mining problems, the scalability associated with the size of the data becomes an important new criterion. According to Witten et al in Anik Andriani, data mining is defined as the process of finding patterns in a sets of data. This process must be in automatic or semi-automatic behavior. The resulting pattern should mean that the pattern gives some advantages. The patterns are to be identified, validated, and later used to make a prediction. From the above notions it can be concluded that Data mining is a process of extracting stored data in large databases [16].

2. Classification

In classification, there are target categories of variables. For example, the classification of income can be separated into three categories: high income, medium income and low income. Then to determine the income of an employee, the classification method is used in data mining.

According to Han et al and Quinlan in their publication paper "Pattern Extraction of On-Time Graduation of STMIK AMIKOM Yogyakarta Student using Data Mining Algorithm C4.5" (Muchamad Widiarto, 2011). Classification techniques in data mining are grouped into decision tree techniques, Bayesian (Naive Bayesian and Bayesian Belief Network), neural networks (Bacpropagation), a technique that is conceptually based on mining association rules and other techniques (K-Nearest Neighboor, genetic algorithm, techniques with rough and fuzzy set approach). Each technique has its own advantages and disadvantages. Data with a particular profile may be most optimal if it is classified by a particular technique, or in other words, a particular data profile might be supported by the advantages of this technique.

3. Decision Tree

Decision tree is one of the most popular classification methods because it is easy to be interpreted by humans. The basic concept of the Decision Tree algorithm is to convert data into decision trees and rules (Defiyanti and Crispina Pardede). The concept of data in the decision tree (Muhammad Syahril, 2011) is represented as in the picture below:

a. Data is expressed in tabular form with attributes and records.

b. The attribute states a parameter created as a criterion in the formation of tree. One of the attributes is an attribute that states the solution data per item called the target attribute.

c. The attribute has values named with instance. For example, the body weight attribute have instances of overweight, average and underweight.

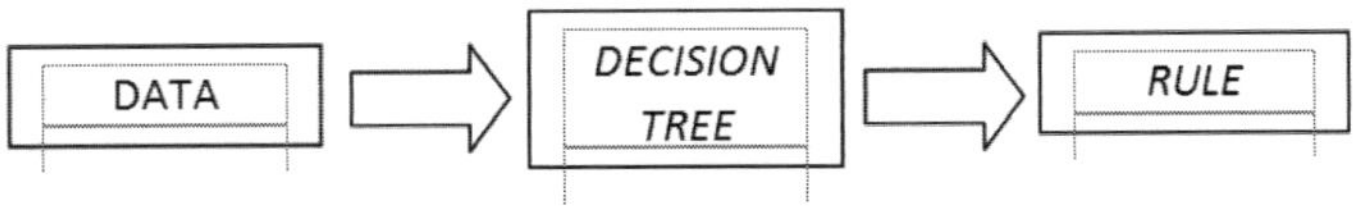

Figure 1. Decision Tree Concept

Decision tree uses a hierarchical structure for supervised learning. The decision tree composition consists of several sections called nodes. The process of the decision tree starts from the root node until the leaf node, which is done recursively, where every branches state a condition that must be met and at the end of a tree declares the class of a data. The decision tree consists of three parts (Ratih and Isye):

a. Root node is the node located at the top of a tree.
b. Internal node is a branching node, there is only one input and has at least two outputs.
c. Leaf node is the final node, has only one input and has no output.

According to Han in Muhammad Syahril (2011), in a decision tree, each leaf node marks the class label. A node that is not the final node consists of a root and an internal node that consists of an attribute test condition on a part of a record that has different characteristics. The root node and internal node are marked with an oval shape and the leaf node is marked with a rectangular shape.

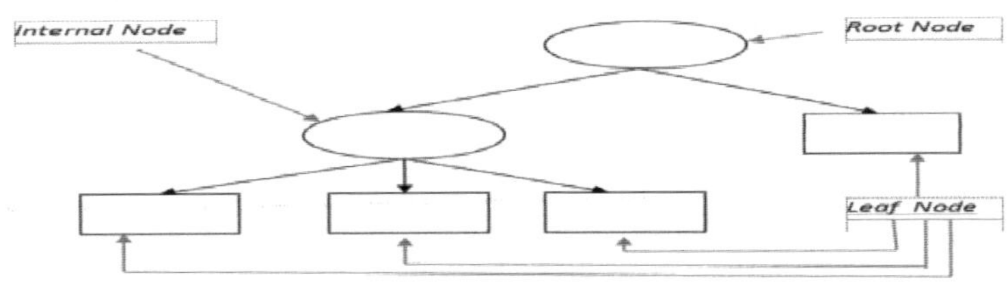

Figure 2. Decision Tree Composition

4. C4.5 Algorithm

According to Sujana in the National Seminar on Information Technology Applications 2010, C4.5 Algorithm is grouped as a decision tree algorithm. This algorithm has input in the form of training samples. Training samples are sample data that will be used to build a tree that has its correctness tested. On the other hand, samples are data fields that will be used as a parameter in doing data classification. The results of the classification process are in the form of rules that later can be used to predict the value of the discrete-type attribute of the new record (Veronica Mortini, 2003). In general, the C4.5 algorithm for building decision trees is as the following (Muhammad Syahril, 2011);

1. Select an attribute as a root.
2. Create a branch for each value.
3. Divide the cases in the branch.
4. Repeat the process for each branch having the same class.

To select an attribute as a root, it is based on the highest gain value of the attributes. To get the value of gain, it must first determine the entropy value.

IOP Conf. Series: Journal of Physics: Conf. Series **970** (2018) 012030 doi:10.1088/1742-6596/970/1/012030

3. Research Methods

The stages in this study follow the general pattern of scientific research as seen in Figure 2.

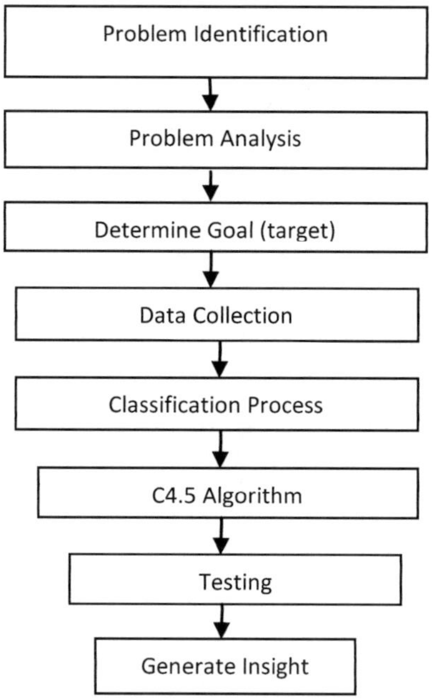

Figure 3. Research Framework

3.1. Research Sites

The research is conducted in the Administrative Office of Labuhan Deli Simpang Jail (Rumah Tahanan). The registration part can be analyzed through custody data with the goal to learn the rules or knowledge, which is the relationship between input and target variables..

3.2. Variables Selection

The data of detainees at the Labuhan Deli State Jail in 2011/2013 have the following format:

1. Registration Number
2. Age
3. Education Level
4. Occupation
5. Address
6. Case

In the classification technique, the output of each targeted data or class must be an integer or a discrete number. From the above detainees data which is used as the target parameter or decision variable (class) is a case containing the value of drug and non-drug parameters. The value of the drug parameter means that the detainee does the type of drug case, while the non-drug parameter value means that the detainee commits another type of crime other than drugs such as theft, robbery, fraud, etc.

IOP Conf. Series: Journal of Physics: Conf. Series **970** (2018) 012030 doi:10.1088/1742-6596/970/1/012030

4. Decision Tree Implementation

Here are the steps to get the Decision Tree model using RapidMiner Decision Tree C4.5 algorithm:

1. Detainees data that contains variables of the conditions and attributes of the decision or target, to know whether the prisoners committed a criminal offense that is type of drug cases and not drugs are stored with Microsoft Excel containing the case to produce the rules information.
2. Prisoner data that has been stored in Microsoft excel can be used directly on the software application Rapid Miner 5.2. By opening the RapidMiner software, then Click New.
3. Then click Repositories, choose Import Excel Sheet. Select the test data saved in Microsoft Excel, click Next.
4. The Data Import Wizard will appear, click Next, on Annotation select Name, click Next to display the Data Import Wizard Step 4.
5. In the number column. select text and id, in the age column select the polynominal and attribute, in the education level column select the polynominal and attribute, in the occupation column select binominal and attribute, in the address column select polynominal and attribute, in the case column as a decision select binominal and label, then click Next.

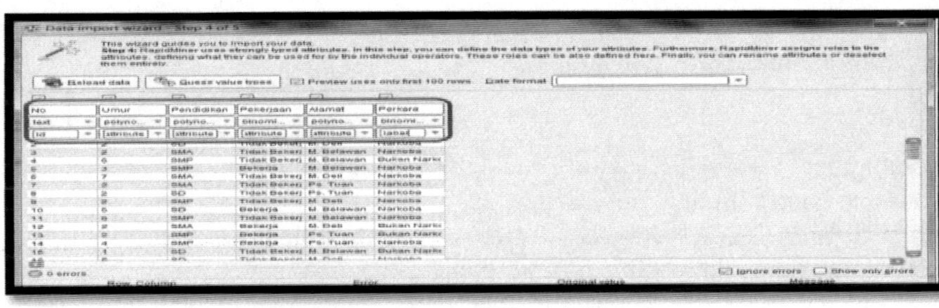

Figure 4. Data Import Wizard Step 4

6. Click Next. Select NewLocalRepository, then click import data. Then click Finish. Then NewProcess RapidMinner will appear.
7. Drag the data on the process screen, then click Operator, Modeling, Classification and Regression, Tree Induction, Decision Tree. Then drag it to the Process screen.
8. Click No Pre Pruning and Pruning, on the criterion select Information_Gain. Click the Run icon to display the decision tree results.

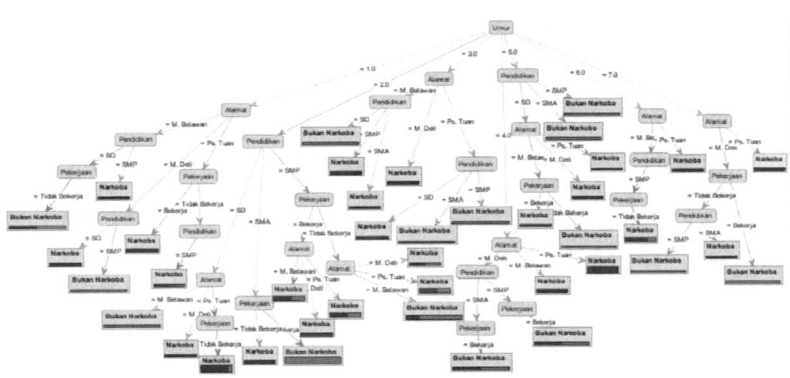

Figure 5. Decision Tree Result

5. Conclusion

From the research that has been done by the author, there are some conclusions that can be drawn and used as knowledge. The conclusions are:

1. In the selection of detainees data variables, which are the conditions attribute and the decisions attributes, they greatly affect the generated rule or knowledge.
2. The more the data being processed, the more rules or knowledge obtained.
3. The decision tree method that is processed by Rapid Miner Software Application is more effectively used in data processing of prisoner at Labuhan Deli Jail.
4. Decision tree of C4.5 Algorithm can provide information or knowledge of a rule that is easily understood by the wider community because it is described with the decision tree.

For further research and development of this application, the author would like to give some suggestions as follows:

1. From the results of rules or knowledge obtained from the data of prisoners, the factors of why the prisoners did criminal acts of drugs can be known. It is expected that the concerning parties, such as the police and the community, can anticipate or reduce the crime of drug offenses.
2. For further development, it would be better to conduct a research combination of several classification techniques to be compared so that it can be known which classification techniques are more effective and efficient

References

[1] Iko Pramudiono. (2003). Pengantar Data Mining : Menambang Permata Pengetahuan di Gunung Data. Kuliah Umum Ilmu Komputer.

[2] Hamidah Jantan1, abdul Razak2. (2010). Human Talent Prediction in HRM Using C4.5 Classification Algorithm. ISSN : 0975-3397.

[3] Angga Ginanjar Mabrur1, Riani Lubis2. (2012). Penerapan Data Mining Untuk Memprediksi Kriteria Nasabah Kredit. Jurnal Komputa Edisi 1 Volume1.

[4] Budianis Dwi Meilani1, Achmad Fauzi Slamat2. (2011). Klasifikasi Data Karyawan Untuk Menentukan Jadwal Kerja menggunakan Metode Decision Tree.

[5] Fajar Astuti Hermawati. (2013). Data Mining. Yogyakarta : Andi.

[6] Anik Andriani. (2012). Penerapan Algoritma C4.5 Pada Program Klasifikasi Mahasiswa Dropout. Seminar Nasional Matematika.

[7] Fernandes Edy Syahputra (2012). Kebijakan Hukum Pidana Terhadap Pengaturan Tindak Pidana Narkotika Di Indonesia. Jurnal Ilmiah.

[8] Angga Ginanjar Mabrur1, Riani Lubis2. (2012). Penerapan Data Mining Untuk Memprediksi Kriteria Nasabah Kredit. Jurnal Komputa Edisi 1 Volume1.

[9] Kusrini dan Luthfi Taufiq Emha. (2009). Algoritma Data Mining. Yogyakarta : Andi.

The development of integrated service centre system for professional teachers empowerment in North Sumatera

S. Gultom, M. M. Simanjorang, Z. Muchtar, A. Mansyur

Mathematics and Science Faculty, Universitas Negeri Medan

syawalgultom@gmail.com, mangaratuasimanjorang@gmail.com, z_muchtar@yahoo.com, abilmansyur@gmail.com

Abstract. Based on Act number 12 in year 2012 the function of higher education is related to individual, social community, knowledge and technology development. Hence, higher education providers need to think and develop policies in order to improve their service and fulfil the higher education function. As part of the effort to fulfil its function *Universitas Negeri Medan (Unimed)*, which historically was a pre-service teacher training institute, should has a special interest on improving teachers' professionalism. The Act number 14 in year 2005 described requirements for professional teacher, including academic qualification and set of competencies possessed by the teacher. The Act also guaranties teachers' right to have opportunities for improving their competencies and academic qualification through training and other professionalism development programme. The question is how this guarantee can be implemented. In order to answer this question a developmental study has been done which aimed on developing an integrated service centre system for professional teachers empowerment. As the name implies, this integrated service centre system is expected to be a real manifestation of Unimed's support towards the improvement of professional teachers quality, which in the end will boils down to the improvement of national education services quality. The result of this study is an integrated service centre system for professional teachers empowerment that fulfils the professionalism principles described in the Act number 14 in year 2005, which has been developed by considering problems faced by and also supports needed by teachers post certification programme.

1. Introduction

Higher education is part of national educational system that has strategic role in creating intelligent citizens and improving science and technology. This strategic role determines competitiveness level of the nation in facing global challenges. Hence, higher education that capable of improving science and technology and producing intellectuals, scientists, and/or professional who are civilized and creative, tolerant, democratic, strong willed, courageous and righteous is important for a nation. The Act of the Republic of Indonesia number 12 of 2012 described that the function of higher education related to individual development, social community and also science and technology. Based on this, a university in Indonesia needs to think some policies that need to be developed for the improvement of university's services and the fulfillment of higher education functions.

Historically *Universitas Negeri Medan (Unimed)* was a pre-service teacher training institute. At that time its services were focused on teacher education. By looking back to this history, it is reasonable if Unimed's role in education especially in North Sumatera province is significance. As part of its effort in performing its function then with its long history in educational world Unimed needs to give special attention in improving teachers' professionalism.

Teachers' professionalism is one of policies taken in this last decade in Indonesia. The Act of Republic of Indonesia number 14 of 2005 described some requirements that have to be fulfilled by a professional teacher including academic qualification and some competences. This Act also guarantee teachers' right to have an opportunity to improve their competences and academic qualification through training and professionalism development programme related to their field of expertise. The question is how to make sure that this warranty is manifested; what kind of policy is needed to support teachers' professionalism; what kind of effort should be taken to improve professional teachers' competences and qualification?

As one of teacher training institutions Unimed needs to think the answers for these questions. Unimed needs to develop a policy that based on current effective regulations to empower techer profession through professional teacher self-development that in the end is expected capable of improving the education and educators quality in Indonesia. In this case, the developed policy should be based on professionalism principles as described in the Act number 14 of 2005 article 7 paragraph 2 that is the self development programme should be democratic, just, non-discriminative, and sustainable while uphold human rights, religious values, cultural values, nation diversity, and professionalism code of ethics.

This article reveals part of the results of an on going study about the development of integrated service center system for professional teacher empowerment. Related to its name this integrated service center system is expected to be a manifestation of Unimed's support for the improvement professional teacher quality that in the end will improve the national educational service quality.

2. Method

This policy study is intended to improve the service available for teacher's professionalism in North Sumatera province. Based on this intention then this study is being held in North Sumatera, which covering teachers with different socio-cultural background.

Complexity of the problem in this study demands a wider view to accommodate high variety information as well. Hence, this study uses a mixed-method approach that involved qualitative and quantitative data. Quantitative data will be collected using a competency test. Data analysis will be enriched with qualitative data, in order to have a comprehensive interpretation and feedback. Hence researchers in this study need to understand different interpretation and different ways in understanding the reality that based on different assumptions. This way the researcher may find complex understanding and make meaning of experiences and others' point of view (Merriam, 2002; Willis, 2007; Taylor, 2008). Different interpretations from different perspectives will enrich the description in this study. Qualitative data is needed for a deeper analysis of a presumption from the quantitative data analysis. Qualitative data are collected through some ways such as questionaire, interview and observation. With these two types of data deeper understanding on the studied problem may be developed.

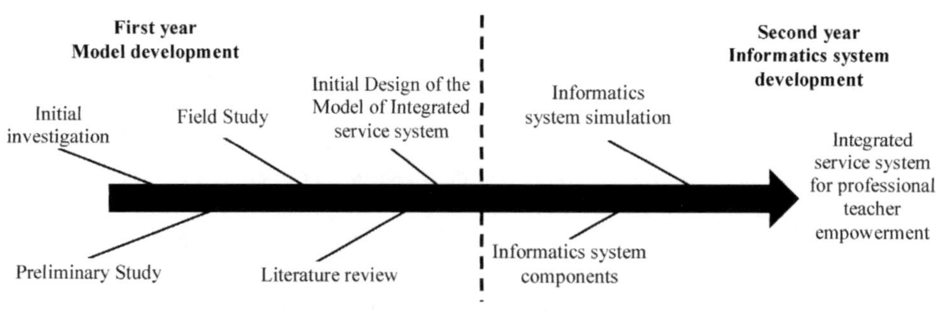

Figure 1. Fishbone chart of research phases

This developmental research covers a wide aspect so that it is necessary to divide this research into two main phases, which are model development and informatics system development. The phases are illustrated by figure 1. Since this study is still on going then the result revealed in this article is not the final result, but an ongoing result especially from the literature review of the current effective regulation in Indonesia related to teacher professionalism.

3. Discussion

This professional teacher empowerment is an opportunity for Unimed to perform its function and responsibility as a teacher training institution. Integrated service system that is being developed in this study is a service policy programme of Unimed to empower professional teachersin order to improve professional teachers' quality that in the end will improve national education quality. This model of integrated service system is constructed by some variables that will be determined by data collected in this study. So far those variable are shown in figure 2.

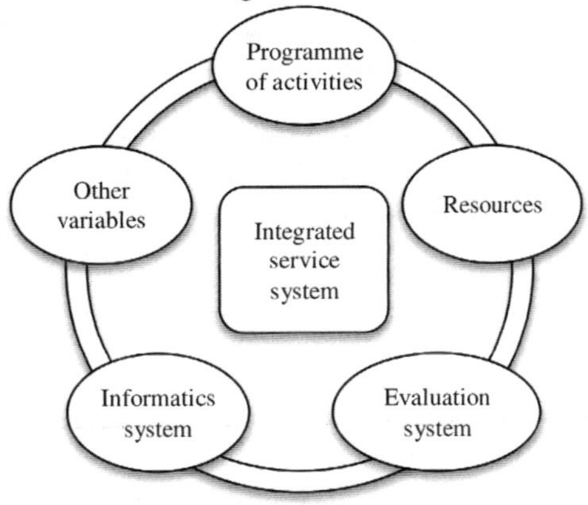

Figure 2. Variables of integrated service center system for professional teacher empowerment

All of these variables may facilitate professional teacher empowerment. They may be explained as follow.

3.1. Service programme

This aspect consists of teacher's competences improvement programme through educational/training programme, such as: training, partnership program, tiered training and special training, short course, advanced education, etc., or non-educational/training programme, such as: discussion about educational problem, seminar, workshop, etc. The service programme is intended to help teacher to obtain the professional teacher competences, which are pedagogic, personal, social and professional competences (Republic of Indonesia Act number 14 of 2005). The programme needs to be filled with transformative education idea (Taylor, 2015), in order to bring transformation in teachers.

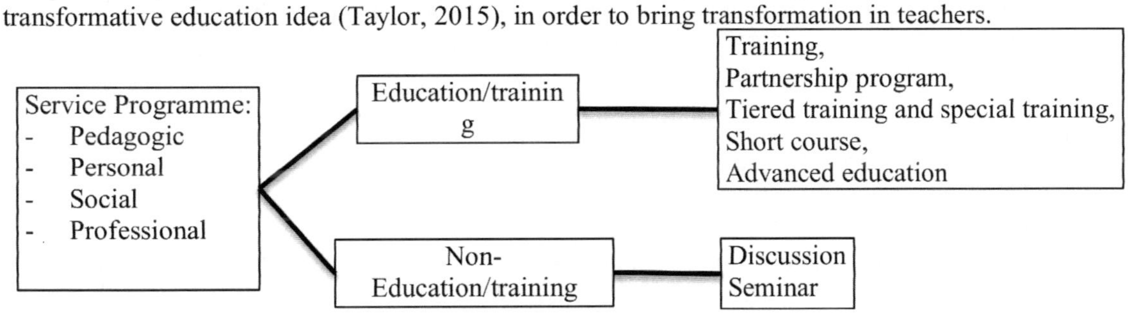

Figure 3. Activities programme

3.2. Resources

In order to guarantee the effectiveness and efficiency of the activities programme, The integrated service system needs some resources, including: human resources experts, tools and media, information about innovation in related theories. According to human resources, Unimed needs to pay more attention in preparing these human resources. The goal of this service is improving teacher professionalism. It is about bringing transformation in teacher, for this purpose human resources provided by Unimed need to understand this transformative education. Experts in this service need to be placed according to his/her expertise. Recruitment system will be developed later for lecturers to be involved in this integrated service system. Human resources in this service will be represent the quality of Unimed and their own, so by being involved in this integrated service system they are promoting Unimed and also being promoted by Unimed. Then it is necessary to maintain the quality standard of human resources in this service system.

3.3. Evaluation System

This evaluation system is intended to evaluate the performance of the system that included effectiveness and efficiency of the system, human resources' performance and also evaluation system for participants who use the service. Regarding participants, since this service about improving teachers' professional competences then interpretation system used in the evaluation should be criterion referenced approach, in order to make sure that participants have mastered the intended competencies. Evaluation system will be designed later after the model of integrated service system is clearer.

3.4. Informatics System

Informatics system will play a role as a tool to integrate the whole system and also to connect Unimed as the service provider to the stakeholders. Hence, it also plays the dissemination role that is disseminating necessary information to those who need it. This informatics system will be developed in the second year of this study.

3.5. Other variables

These variables may be clearer by the end of this model development process after collecting the necessary data. Researchers include this variable to accommodate other findings related to the integrated service system model that may be found later in the process of this study.

4. Conclusion

The integrated service system is intended to improve teachers' professional competencies. This service system is developed to support professional teacher empowerment policy, which is regulated in current regulation in Indonesia about teacher professionalism. This developmental study is still ongoing. So far there are four variables that construct the integrated service system model, they are service programme, resources, evaluation system and informatics system. This system will be an indicator of Unimed's commitment in empowering teachers' professionalism and improving educational practices in North Sumatera and in Indonesia.

References

[1] Merriam, S. B. (2002). Introduction to qualitative inquiry. In A. S.B. Merriam (Ed.), *Qualitative research in practice: Example for discussion and analysis* (pp. 3- 17). San Fransisco: John Wiley & Sons, Inc.

[2] Taylor, P. C. (2008). Multi-paradigmatic research design spaces for cultural studies researchers embodying postcolonial theorising. *Culture Studies of Science Education*, 3(4), 881–890.

[3] Taylor, P. C. (2015). Constructivism. In R. Gunstone (Ed.), *Encyclopedia of Science Education* (pp. 218-224). Dordrecht, The Netherlands: Springer.

[4] __________. (2015). Transformative Science Education. In R. Gunstone (Ed.), *Encyclopedia of Science Education* (pp. 1079-1082). Dordrecht, The Netherlands: Springer.
[5] The Act of the Republic of Indonesia number 12 of 2012
[6] The Act of the Republic of Indonesia number 14 of 2005
[7] Willis, J. W. (2007). *Foundation of qualitative research: Interpretive and critical approaches*. Thousand Oaks: Sage Publications.

Biology Procedural Knowledge at Eleventh Grade of Senior High School in West Lampung Based on Curriculum

T M Sari[1], Paidi[2], I S Mercuriani[2]

[1]Post-graduate Students of Biology Education, Yogyakarta State University, Yogyakarta
[2]Biology Education Department, Yogyakarta State University, Yogyakarta, Indonesia

[1] corresponding author: tika.mayang30@gmail.com

Abstract This study was aim to determine Biology procedural knowledge of senior high school in West Lampung based on curriculum at 11th grade in even semester. This research was descriptive research. The population was all students of senior high school in West Lampung. The sampling technique in this research used purposive sampling technique, so the researcher obtained 3 schools using K13 and 3 schools using KTSP. Data collecting technique used instrument test. Data analysis technique used U-Mann Whitney test. The result showed that $p=0.028$ ($p<0.05$), so there was significant differences between school using K13 and KTSP. The procedural knowledge of schools which using K13 is higher than school which using KTSP, with the mean score K13=4.35 and KTSP=4.00

1. Introduction

Education Unit Level Curriculum (KTSP) and 2013 Curriculum which emphasizes the development of students' thinking skills, as one aspect of competence, have been used for several years in schools, including biology subject in high school. Standards the education assessment includes 3 domains: Affective domain, Cognitive domain, and Psychomotor domain. In the cognitive domain has the ability of processes and products. In the cognitive product dimension (knowledge) consists of (1) factual knowledge, (2) conceptual knowledge, (3) procedural knowledge, (4) and metacognitive knowledge. While the cognitive process dimension includes six process categories (C1-C6) that is (1) remembering (2) understanding (3) applying (4) analyzing (5) evaluating (6) and creating.

Permendikud (The Ministry of Education and Culture) No. 20 2016 concerning Graduates Competency Standards Elementary and High School clearly states the need for high school students is trained in cognitive processes that includes six categories of process (C1-C6) from remembering, understanding, applying, analyzing, and evaluating to achieve mastery of conceptual knowledge (conceptual knowledge), procedural knowledge (knowledge), and metacognitive knowledge (metacognitive knowledge).

The Government through the Ministry of Education and Culture in 2013 began implementing the change of replacement curriculum from the Education Unit Level Curriculum (KTSP) to a new curriculum called 2013 Curriculum. The 2013 curriculum which based on character and competence was born as the answer of various criticisms of the 2006 curriculum. 2013 Curriculum Development is a continuity of steps that include the competence of attitudes, knowledge, and skills in an integrated manner. Biology which related to the nature, the students are emphasized to know about nature, therefore K13 uses a scientific approach.

By the turning of KTSP into the 2013 curriculum the schools are expected there is increasing in the education quality that directly impact on improving learners in Learning process. The curriculum should not be viewed as an activity, but directly focused on the expected learning outcomes (Intended learning outcome). If every turn of the curriculum aims to provide better learning for learners, one of them for the process and cognitive products of students is to be better. However, not all teacher or lecturers are ready to turn of the curriculum, the example of the change of 2013 curriculum there are still schools that use the KTSP curriculum. High school in West Lampung district, not all of them simultaneously using curriculum 2013, there are still some schools that use KTSP curriculum. From 12 high school (SMA) Negeri in West Lampung, only 3 schools are implementing the 2013 curriculum.

2. Material and Method

2.1 Material

2.1.1 Procedural Knowledge

Procedural knowledge is the knowledge of how to do something. Doing this may be doing routine exercises up to solving new problems. Procedural knowledge is a step that must be followed. This knowledge includes the knowledge of skills, algorithms, techniques, and methods, which of all are referred to as procedures. Procedural knowledge also includes the criteria for determining the procedure to be performed (Anderson and Krathwohl, 2001).

Table. 1 Educational Taxonomy Revision

Knowledge Dimension	Cognitive Process Dimension					
	Remember	Understand	Aplication	Analysis	Evaluation	Create
Factual						
Conceptual						
Prosedural						
Metacognitive						

Conceptual knowledge, procedural knowledge, and procedural flexibility were assesed reliably and partly independen of each other. Conceptual knowledge may hep with the construction, selection, and appropriate execution of problem solving procedureat the same time, practice using procedure may help students develop and deepen understanding of concepts (Schneider et al, 2011)

2.1.2 Curriculum

The meaning of the curriculum in the juridical-formal perspective, that is (UU) no. 20 of 2003 on National Education System. "The curriculum is a set of plans and arrangements concerning of objectives, content, materials and ways to used as guidelines for the implementation of learning activities to achieve educational goals". The 2013 curriculum is a refinement of the Education Unit Level Curriculum (KTSP). The objective of the 2013 Curriculum is to prepare Indonesian people to have the ability for live as individuals and citizens who are faithful, productive, creative, innovative, and affective and able to contribute to the life of society, nation, state and civilization of the world.

2.2 Method

This research used descriptive method research. This research was conducted in West Lampung. Data collection was conducted on May 4 to May 23, 2017 in the even semester of academic year 2016/2017. The sampling technique used is purposive sampling. The sample in this research was the students of XI grade of Senior high school in West Lampung which 3 schools that use K13 and 3 schools still use KTSP. The number of sample is 348 students.

Data was collected through the test technique. The test technique has been done to measure the achievement of procedural capability. The instrument test that was used is multiple choice and essay consisting of immune system material, reproductive system, coordination system, and excretion system. The instrument has been compiled and validated. Analytical techniques was used are inferential analysis in the form of different test to know the difference of students' procedural knowledge achievement using K13 and KTSP

3. Result and Discusion

3.1 Result

The results of this research are the result of an instrument in the form of multiple choice and essay of XI grade of senior high school on material which consists of immune system material, reproduction system, coordination system, and excretion system. Based on the result, the researcher obtained p value = 0,028 ($p<0,05$), so there are significant difference between school using K13 and KTSP. The average value of the procedural knowledge students' achievement of XI grade of science class in West Lampung which use K13 is 4.35 and school which use KTSP got an average value is 4.00.

The comparison of procedural knowledge value which associated with the curriculum that has been used is shown in Table 2 below.

Table 2. The Procedural Knowledge Value with Different Curriculum

Variabel	Statistics (Mean)	Sig (Mann Whitney Test)
K13	4.35	Significant differences (p = 0.028)
KTSP	4.00	

3.2 Discusion

Anderson and Kratwohl (2001) divide students cognitive abilities into two, cognitive processes and products. This cognitive process consists of C1-C6 that is to remember, understand, apply, analyze, evaluate, and create, and the cognitive product consists of factual, conceptual and procedural.

There are differences in the school outcomes that use 13 curriculum and KTSP in West Lampung. This can be seen from the value of p = 0.028 ($p <0.05$) which schools use the 2013 curriculum get a higher average score for about 4.35 while schools that use KTSP is 4.00. The 13 curriculum through the scientific approach is to suppress the students to be able to know the procedures to do scientific things in biology subject. There is a difference between the two curriculums.

This result is supported by the research that has been conducted by Domili (2015) there is a difference in learning outcomes between schools use 13 curriculum obtaining an average (80.04) while schools use KTSP earn an average (77.08). The research that conducted by Tanaya (2017) The quality and quantity of students' questions at XI grade MIA SMA Negeri 1 Madiun identified that the quality of students' questions C1 is 40% conceptual dimension, C2 level 10% conceptual dimension, C2 level procedural dimension 30%, level C2 dimension metacognition 20 %.

Engelbrecht et. al (2017) state the balanced view, on the other hand, considers both conceptual and procedural skills as essential in engineering education and job through three strands; in linking its conceptual aspect as a basis on which the conceptual can be built, and in facilitating the progression from the conceptual problem to its procedural solution. There is a complex relationship between conceptual and procedural knowledge in mathematics, dynamic in character. This relates to both, how to understand the constructs and the implications these analytic categories in the organisation of teaching, as well as how engineering students view their relevance for their education.

Procedural knowledge is required in learning biology, as we know biology is closely related to nature and organism, so it is necessary students' understanding about the way or the procedure, the steps to do something in learning biology.

4. Conclusion

Based on the results of the procedural knowledge achievement on the learning Biology at XI grade for some senior high schools in West Lampung shows that the curriculum was used in the schools affects the students' procedural knowledge, in K13 get a better value than the school which uses the curriculum 13. The ability of students' procedural knowledge in West Lampung is still low both using K13 and KTSP. It is expected that the teacher and facilitator can improve the quality of education nowdays.

Refferences

[1] Anderson L.R, Krathwohl D.R 2001 *A Taxonomy for Learning, Teaching and Assessing: A Revision of Bloom's Taxonomy of Educational Objectives*. A Bridged Edition. New York: Longman.

[2] Departmen Pendidikan Nasional 2003 *Undang-undang Nomor 20 tahun 2003,* Tentang Sistem Pendidikan Nasional Jakarta Depdiknas

[3] Domili Z, Ilham A, and Halidu S, 2015 *Perbedaan Hasil Belajar Siswa dengan Menerapkan Kurikulum KTSP dan Kurikulum 2013 di kelas IV MI-Al Mourku Kecamatan Telaga Kabupaten Gorontalo* thesis ung

[4] Engelbrecht J, Bergsten C, and Kågesten 2017 *Conceptual and Procedural approaches to mathematics in the engineering curriculum: view of qualified engineers*, (European Journal of Engineering Education) vol 42 (5) p 13

[5] Permendikbud 2016 *Peraturan Menteri Pendidikan dan Kebudayaan Nomor 20 Tahun 2016 tentang SKL Pendidikan dasar dan Menengah* Jakarta

[6] Schneider, M., Rittle-Johnson, B., & Star, J. R. 2011. *Relations among conceptual knowledge, procedural knowledge, and procedural flexibility in two samples differing in prior knowledge. Developmental psychology*, vol *47* (6), 1532.

[7] Tanaya N.Y.T.B, Suciati, and Maridi 2017 *Profil Kualitas dan Kuantitas Pertanyaan Siswa Pada Pembelajar Biologi kelas XI MIA SMA Negeri Madiun,* p 81